HEINEMANN MODULAR MATHEMATICS for EDEXCEL AS AND A-LEVEL
Revise for Statistics 3

Greg Attwood Gordon Skipworth Gill Dyer

1 Combinations of random variables — 1
2 Sampling — 5
3A Estimation and confidence intervals — 11
3B Hypothesis tests for means — 18
4A Goodness of fit — 26
4B Contingency tables — 36
5 Correlation — 42
6 Projects — 50

Examination style paper — 55

Answers — 58

Heinemann Educational Publishers,
a division of Heinemann Publishers (Oxford) Ltd,
Halley Court, Jordan Hill, Oxford, OX2 8EJ

OXFORD MELBOURNE AUCKLAND JOHANNESBURG
BLANTYRE GABORONE PORTSMOUTH NH (USA) CHICAGO

© Greg Attwood, Gordon Skipworth and Gill Dyer 2001

All rights reserved. No part of this publication may be reproduced, stored in a retrieval system, or transmitted in any form or by any means, electronic, mechanical, photocopying, recording, or otherwise without either the prior written permission of the Publishers or a licence permitting restricted copying in the United Kingdom issued by the Copyright Licensing Agency Ltd, 90 Tottenham Court Road, London W1P 9HE.

First published 2001

05 04 03 02 01
10 9 8 7 6 5 4 3 2 1

ISBN 0 435 51118 1

Cover design by Gecko Limited

Original design by Geoffrey Wadsley; additional design work by Jim Turner

Typeset and illustrated by Tech-Set Limited, Gateshead, Tyne and Wear

Printed in Great Britain by Scotprint

Acknowledgements:

The publisher's and authors' thanks are due to Edexcel for permission to reproduce questions from past examination papers. These are marked with an [E].

The answers have been provided by the authors and are not the responsibility of the examining board.

About this book

This book is designed to help you get your best possible grade in your Statistics 3 examination. The authors are Chief and Principal examiners and moderators, and have a good understanding of Edexcel's requirements.

Revise for Statistics 3 covers the key topics that are tested in the Statistics 3 exam paper. You can use this book to help you revise at the end of your course, or you can use it throughout your course alongside the course textbook, *Heinemann Modular Mathematics for Edexcel AS and A-level Statistics 3*, which provides complete coverage of the syllabus.

Helping you prepare for your exam

To help you prepare, each topic offers you:

- **Key points to remember** – summarise the statistical ideas you need to know and be able to use.
- **Worked examples and examination questions** – help you understand and remember important methods, and show you how to set out your answers clearly.
- **Revision exercises** – help you practise using these important methods to solve problems. Past paper questions are included so you can be sure you are reaching the right standard, and answers are given at the back of the book so you can assess your progress.
- **Test yourself questions** – help you see where you need extra revision and practice. If you do need extra help they show you where to look in the *Heinemann Modular Mathematics for Edexcel AS and A-level Statistics 3* textbook.

Exam practice and advice on revising

Examination style practice paper – this paper at the end of the book provides a set of questions of examination standard. It gives you an opportunity to practise taking a complete exam before you meet the real thing. The answers are given at the back of the book.

How to revise – for advice on revising before the exam, read the How to revise section on the next page.

How to revise using this book

Making the best use of your revision time

The topics in this book have been arranged in a logical sequence so you can work your way through them from beginning to end. But **how** you work on them depends on how much time there is between now and your examination.

If you have plenty of time before the exam then you can **work through each topic in turn**, covering the key points and worked examples before doing the revision exercises and Test yourself questions.

If you are short of time then you can **work through the Test yourself sections** first, to help you see which topics you need to do further work on.

However much time you have to revise, make sure you break your revision into short blocks of about 40 minutes, separated by five- or ten-minute breaks. Nobody can study effectively for hours without a break.

Using the Test yourself sections

Each Test yourself section provides a set of key questions. Try each question:

- If you can do it and get the correct answer then move on to the next topic. Come back to this topic later to consolidate your knowledge and understanding by working through the key points, worked examples and revision exercises.
- If you cannot do the question, or get an incorrect answer or part answer, then work through the key points, worked examples and revision exercises before trying the Test yourself questions again. If you need more help, the cross-references beside each Test yourself question show you where to find relevant information in the *Heinemann Modular Mathematics for Edexcel AS and A-level Statistics 3* textbook.

Reviewing the key points

Most of the key points are straightforward ideas that you can learn: try to understand each one. Imagine explaining each idea to a friend in your own words, and say it out loud as you do so. This is a better way of making the ideas stick than just reading them silently from the page.

As you work through the book, remember to go back over key points from earlier topics at least once a week. This will help you to remember them in the exam.

Combinations of random variables

Key points to remember

1 **Distribution of a linear combination of random variables**
If X_1 and X_2 are independent random variables such that $X_1 \sim N(\mu_1, \sigma_1^2)$ and $X_2 \sim N(\mu_2, \sigma_2^2)$ then
$$X_1 \pm X_2 \sim N(\mu_1 \pm \mu_2, \sigma_1^2 + \sigma_2^2)$$

Note: you always use $\sigma_1^2 + \sigma_2^2$ regardless of whether you are considering $(X_1 + X_2)$ or $(X_1 - X_2)$.

2 **Distribution of a linear combination of random variables**
If X_1 and X_2 are independent random variables such that $X_1 \sim N(\mu_1, \sigma_1^2)$ and $X_2 \sim N(\mu_2, \sigma_2^2)$ then
$$aX_1 \pm bX_2 \sim N(a\mu_1 \pm b\mu_2, a^2\sigma_1^2 + b^2\sigma_2^2)$$

You always use $a^2\sigma_1^2 + b^2\sigma_2^2$ regardless of whether you are considering $(aX_1 + bX_2)$ or $(aX_1 - bX_2)$.

3 **Extension to more random variables**
The above may be extended to cover as many random variables as you want.
e.g. If X_1 and X_2 are as described above and $X_3 \sim N(\mu_3, \sigma_3^2)$ then
$$Y = X_1 - X_2 + 3X_3 \sim N(\mu_1 - \mu_2 + 3\mu_3, \sigma_1^2 + \sigma_2^2 + 9\sigma_3^2)$$

Example 1

If $X_1 \sim N(15, 3^2)$, $X_2 \sim N(10, 2^2)$ and $X_3 \sim N(12, 4^2)$ find the distribution of Y where Y is defined by
$$Y = 2X_1 - X_2 + 4X_3$$

Answer

$Y = 2X_1 - X_2 + 4X_3$
so $Y \sim N([2 \times 15] - 10 + [4 \times 12], [4 \times 3^2] + 2^2 + [16 \times 4^2])$ Using **3**
$\sim N(68, 296)$

Example 2

The weight of wine in a 75 cl bottle is normally distributed with mean 752 grams and standard deviation 4 grams. The distribution of the weight of the bottles is normal with mean 420 grams and standard deviation 12 grams.

2 Combinations of random variables

Twelve bottles at a time are stored in cardboard cartons. The weight of the cartons is normally distributed with mean 602 grams and standard deviation 2.5 grams.
If X represents the weight of a carton with 12 full bottles, find the distribution of X.

Answer

Let W represent the weight of wine in a bottle, then $W \sim N(752, 4^2)$.
Let B represent the weight of a wine bottle, then $B \sim N(420, 12^2)$.
Let C represent the weight of a carton, then $C \sim N(602, 2.5^2)$.
$X = (W_1 + W_2 + \ldots + W_{12}) + (B_1 + B_2 + \ldots + B_{12}) + C$
so $X \sim N([12 \times 752] + [12 \times 420] + 602, [12 \times 16] + [12 \times 144] + 6.25)$ Using **1**
$\therefore X \sim N(14\,666, 1926.25)$

Worked examination question 1[E]
The random variable S is defined as

$$S = W + P + 3C$$

where $W \sim N(10, 4^2)$, $P \sim N(15, 5^2)$, $C \sim N(12, 2^2)$, and W, P and C are independent.
Find:
(a) $E(S)$
(b) $Var(S)$
(c) $P(S < 66)$

Random variables C_1, C_2, and C_3 are independent and each has the same distribution as C. The random variable T is defined as

$$T = W + P + \sum_{i=1}^{3} C_i$$

(d) Find $E(T)$
(e) Show that $Var(S) - Var(T) = 24$.

Answer
(a) $E(S) = E(W) + E(P) + 3E(C)$ Using **2**
$= 10 + 15 + (3 \times 12)$
$= 61$

(b) $Var(S) = Var(W) + Var(P) + 9Var(C)$ Using **2**
$= 16 + 25 + (9 \times 4)$
$= 77$

(c) $P(S < 66) = P\left(Z < \dfrac{66 - 61}{\sqrt{77}}\right)$
$= P(Z < 0.57)$
$= 0.7157$

(d) $E(T) = E(S) = 61$ Using **1**

(e) $\text{Var}(T) = \text{Var}(W) + \text{Var}(P) + 3\text{Var}(C)$
    $= 16 + 25 + (3 \times 4)$
    $= 53$

$\therefore \text{Var}(S) - \text{Var}(T) = 77 - 53 = 24$

Using **1**

Revision exercise 1

1 $X \sim N(20, 4^2)$ and $Y \sim N(30, 3^2)$:
   Find the distribution of:
   (a) W, where $W = 2X + Y$,
   (b) Y, where $Y = X - 3Y$.

2 If X_1, X_2 and X_3 are independent random variables with $X_1 \sim N(4, 3)$, $X_2 \sim N(5, 2)$ and $X_3 \sim N(10, 4)$ find the distribution of Y where:
   (a) $Y = X_1 - X_2 + X_3$,
   (b) $Y = X_1 + X_2 + X_3$,
   (c) $Y = 3X_1 + 2X_2 + X_3$,
   (d) $Y = 4X_1 - 2X_2 + 3X_3$.

3 The weights of stock cubes are normally distributed with mean 6.36 grams and standard deviation 0.25 grams.
   The cubes are packed together in twelves in cardboard packets. The weight of a packet is normally distributed with mean 5 grams and standard deviation 0.1 grams.
   20 packets are in turn gathered and shrink-wrapped together. The weight of the shrink-wrapping is normally distributed with mean 2 grams and standard deviation 0.1.
   If X represents the weight of 20 shrink-wrapped packets, find the distribution of X.

4 The random variables X_1, X_2, X_3 and X_4, are independent, each with mean μ and variance σ^2.
   The random variable Z is defined as $Z = X_1 + \ldots + X_4$, and the random variable Y is defined as $Y = X_1 + 3X_4$.
   (a) Find the ratio of $\text{Var}(Z)$ to $\text{Var}(Y)$.
   A certain brand of biscuit is individually wrapped. The weight of a biscuit can be taken to be normally distributed with mean 75 g and standard deviation 5 g.

4 Combinations of random variables

The weight of an individual wrapping is normally distributed with mean 10 g and standard deviation 2 g. Six of these individually wrapped biscuits are then packed together. The weight of the packing material is a normal random variable with mean 40 g and standard deviation 3 g.
(b) Find the probability that the total weight of the packet lies between 535 g and 565 g. [E]

Test yourself	What to review
	If your answer is incorrect:
1 The random variables X_1 and X_2 are independent with $X_1 \sim N(100, 6^2)$ and $X_2 \sim N(150, 5^2)$. **(a)** Find the distribution of Y if $Y = X_1 + X_2$. **(b)** Find the distribution of Z if $Z = X_2 - X_1$.	*Review Heinemann Book S3 pages 1–6*
2 X and Y are independent normal variables with $X \sim N(2, 2)$ and $Y \sim N(6, 3)$. **(a)** Find the distribution of W where $W = 3X + 2Y$. **(b)** Find the distribution of Z where $Z = 3Y - 2X$.	*Review Heinemann Book S3 pages 1–6*
3 The Yummy biscuit company manufactures biscuits in presentation boxes. Each box contains 12 foil-wrapped chocolate biscuits and 12 unwrapped plain biscuits. The weights of chocolate biscuits in their wrappers are normally distributed with mean weight 16 grams and standard deviation 0.15 grams. The weights of the plain biscuits are normally distributed with mean weight 12 grams and standard deviation 0.15 grams. The weights of the boxes themselves are normally distributed with mean weight 12 grams and standard deviation 0.3 grams. **(a)** Find the distribution of the weights of the presentation boxes complete with biscuits. **(b)** Find the probability that a presentation box of biscuits has a weight > 350 grams.	*Review Heinemann Book S3 pages 1–6*

Test yourself answers

1 (a) N(250, 61) **(b)** N(50, 61)   **2 (a)** N(18, 30) **(b)** N(14, 35)   **3 (a)** N(348, 0.63) **(b)** 0.006

Sampling

Key points to remember

1. **Simple random sampling**
   A sample of size n is called a simple random sample if every possible sample of size n taken from a population of size N, has an equal chance of being selected.

   Advantage: It is free from bias.
   Disadvantage: Not suitable for large populations.

2. **Random number sampling**
   A method in which each element of the population is assigned a number and then random number tables (or other random number generators) are used to select numbers and thus the corresponding sampling units.

   Advantages: Free from bias; easy to use with each number having a known equal chance of selection.
   Disadvantages: Not suitable for large populations.

3. **Systematic sampling**
   A method in which a sample is obtained by choosing elements at regular intervals from an ordered list.

   Advantages: It is easy to use; suitable for large sample sizes.
   Disadvantages: An ordered list is needed; it is only random if the ordered list is random.

4. **Stratified sampling**
   The population is divided into mutually exclusive groups (strata), and a simple random sample is taken from each stratum in proportion to its size.

   To use stratified sampling, clear strata must be identified.

   Advantages: Can give more accurate estimates than simple random sampling when clear strata are present; gives separate estimates for each stratum.
   Disadvantages: Strata must be distinct, have means as different as possible, and have small variances.

5 **Quota sampling**
A non-random sampling method in which an interviewer selects people and fits them into groups (or quotas) according to specific characteristics. The characteristics and the number in each quota are selected beforehand so that the quotas reflect the known population characteristics.
Advantages: Enables field work to be done quickly; cheap; easy to administer.
Disadvantages: Not possible to estimate sampling errors; the interviewer may not be able to judge the characteristics easily; non-responses not recorded.

Example 1
A student wishes to investigate the relationship between first and second names within the sixth form at her school. There are 200 students in the sixth form, and she only wishes to take a sample of 50. She decides to number the students from 0 to 199 and to use random number tables to select her sample. Beginning at the 1st column and 5th row in Table 7 of random numbers in Book S3 with the numbers 99, 09 she reads horizontally.

> You could use the numbers 1 to 200.
>
> See **2**

(a) List the next 6 numbers she selected.
(b) What is wrong with her proposed method, and how would you recommend the sample should be selected?

Answer
(a) The next 6 numbers are 39, 25, 66, 31, 70 and 56.
(b) Numbers greater than 99 would not be selected.
She should group the numbers in threes so the numbers would read 990, 939, 256, 631, 705, etc. To save discarding numbers higher than 199 she could assign 5 numbers to each student, so the first student would have the numbers 0, 200, 400, 600 and 800 while the last would have the numbers 199, 399, 599, 799 and 999.

> Using Table 7 in Book S3.
>
> If you use the numbers 1 to 200 then either you have to assign 4 numbers or you have to use 0 instead of 1000.

Example 2
A film company wishes to publish the number of viewers for each of its films shown on television. They intend to send interviewers on to the streets of large cities to ask people which films they watched during the past week.
(a) Suggest a suitable sampling method.
(b) What are the advantages and disadvantages of this method?

Answer
(a) A quota sampling method could be used.
(b) The advantages of this method are that it is quick, cheap and easy to administer.
   The disadvantages are that it is non-random, it may be difficult to judge the characteristics, and non-responses are not recorded.

> Using **5**

Example 3
In a mixed upper school there are a total of 800 pupils in the three upper years. The final year has only 200 pupils, the others have 300 each. A survey is to be carried out to discover the children's favourite types of television programme. It is thought that there will be a difference between year groups because of the different ages of the pupils.
(a) Suggest a suitable method of selecting pupils to fill in a questionnaire.
(b) If a sample of 160 is to be sampled, explain how the individual pupils might be selected in order to get a representative sample.

Answer
(a) The pupils fall into separate, non-overlapping strata years, so a stratified sample should be taken.

Using **4**

(b) A random sample of 40 is taken from the final year and random samples of 60 from each of the other years. The pupils in each year group are numbered and random number tables used to select pupils from each year.

Worked examination question 1 [E]
There are 12 six-person tents at an army cadet camp, all fully occupied. Two of these are for female cadets and the other 10 are for male cadets. The camp leader wishes to estimate the average amounts spent by each cadet on kit and instructs his assistant to take a sample of 18 cadets to obtain an estimate.
(a) Suggest an appropriate sampling technique that the assistant should use.
(b) Advise the assistant how to go about obtaining this sample.

Answer
(a) Since there are clear strata consisting of unequal numbers of males and females, a stratified sampling system should be used.

Using **4**

(b) $\frac{5}{6}$ of the cadets are males, so the number of males selected should be $\frac{5}{6}$ of the total.

$$\frac{5}{6} \times 18 = 15$$

The assistant should select at random 3 female cadets and 15 male cadets.

Worked examination question 2 [E]
The 240 members of a bowling club are listed alphabetically in the club's membership book. The committee wishes to select a sample of 30 members to fill in a questionnaire about the facilities the club offers.
(a) Explain how the committee could use a table of random numbers to take a systematic sample.
(b) Give one advantage of this method over taking a simple random sample.

8 Sampling

Answer

Using ▇2 and ▇3,
(a) The members would be labelled 1 → 240.
The random number table would be used to select a number from 1 to 8. Starting from this number, every eighth number would be selected and hence its corresponding member (e.g. 6, 14, 22, ...).
(b) It is easier to use.

> The number is selected from 1 to 8 because the sample is $\frac{30}{240} = \frac{1}{8}$ of members.

> There are other advantages but these are shared by simple random sampling.

Revision exercise 2

1. A researcher employed by a television company needs to sample attitudes to work in a large company. The company has three factories, one in Birmingham, one in Aberdeen and one in Cardiff, employing in all about 2600 people. To collect data, the researcher plans to distribute questionnaires to 10 employees. He intends to distribute the questionnaires to the first 10 workers to leave the factory in Birmingham after the night shift on the following day (a Monday).
   (a) Give *three* reasons why this strategy is likely to produce a biased response.

   The company is prepared to supply the researcher with a list of employees in each factory.
   (b) Outline *briefly* how the researcher could select a systematic sample.
   (c) Outline *briefly* how the researcher could select a stratified sample. [E]

2. Using Table 7 of random numbers in Book S3, starting at the 5th row and working across, 4 two-digit random numbers between 01 and 15 were selected. The first number was 09.
   (a) Write down the other three numbers.
   (b) Give an example of a population for which it would be suitable for these 4 numbers to be used to select a random sample of size 4 without replacement. You should specify the population size and the sampling frame you would use.

3. A researcher has the task of telephoning 60 telephone subscribers in order to ask them questions for a magazine article. He is asked to choose them randomly from those listed in the non-business section of the local telephone directory.

The non-business pages start on page number 213 and run to page number 435, inclusive. On each page of this section there are 4 columns, each with 70 names.

He decides to make each selection by the following method. He will choose a page at random, and then choose a column at random from the selected page. He will then take the top subscriber in that column, unless he has already used that subscriber. If the top subscriber has been used already he will use the first subscriber in that column who has not yet been used.

(a) State clearly how he can use Table 7 of random numbers in Book S3 in order to choose the pages and the columns.

(b) Give a reason why his method will not produce a simple random sample of all the subscribers listed in the telephone directory. [E]

4 A researcher from an agricultural research station wished to estimate the yield from a rectangular field of barley that was 140 m by 60 m. The field was divided into 84 squares, each 10 m × 10 m, and each square was assigned a number from 01 to 84.

Using Table 7 in Book S3, starting from the beginning of the 6th row and working across the row, the researcher selected a simple random sample of 7 squares. The first two squares were those numbered 56 and 32.

(a) Find the numbers of the other 5 squares.

The researcher noticed that 4 of these squares were from the edges of the field close to the hedgerow but only 36 of the original 84 squares were along the edge. Squares along the edge of the field, and therefore closer to the hedgerow, will have a lower yield. The researcher therefore decided to take a different sample of size 7 that would reflect the fact that only 36 of the 84 squares lay along the edge of the field.

(b) Suggest a suitable sampling method.

(c) Explain how the researcher could take a sample of size 7 using this sampling method. [E]

Sampling

Test yourself	What to review
	If your answer is incorrect:
1 Explain what is meant by a simple random sample of size n.	*Review Heinemann Book S3 page 8*
2 A simple random sample of size 8 of the numbers between 0 and 50 inclusive was taken without replacement. Using the random number Table 7 in Book S3, starting at the top of the 7th column with the number 39 and working downward, the first two numbers were 09 and 01. **(a)** Find the other 6 numbers. **(b)** Explain how these numbers could be used to take a random sample of size 8.	*Review Heinemann Book S3 pages 8–10*
3 An engineering firm with 2000 employees wishes to give questionnaires to 40 of them in order to get the views of the workforce on a possible change to the working system. Each of the employees can be identified by their employee number. It is suggested by one manager that the lowest 40 numbers should be used since these belong to the group of workers who have worked longest for the firm. **(a)** What is wrong with this method of distributing the questionnaires? **(b)** Suggest a more suitable method.	*Review Heinemann Book S3 pages 8–16*
4 A survey is to be undertaken at a junior school to determine what types of books are read most by children in the age range 7 to 11. The pupils are divided into 4 year groups according to age. **(a)** Suggest a suitable method for selecting a sample of pupils' opinions. **(b)** Give the advantages and disadvantages of this method.	*Review Heinemann Book S3 pages 11–12*
5 (a) Describe what is meant by quota sampling. **(b)** Give an instance of when it might be used.	*Review Heinemann Book S3 pages 12–14*

Test yourself answers

1 A sample of size n is called a simple random sample if every possible sample of size n has an equal chance of being selected from a population of size N.
2 **(a)** 18, 27, 35, 13, 05, 46 **(b)** The units of the population would be numbered and those corresponding to the numbers in part (a) would be selected.
3 **(a)** This would be a non-random sample as only the views of the longest serving employees would be found.
  (b) A more suitable method would be to use systematic sampling.
4 **(a)** Stratified sampling would be more suitable because their reading preferences might vary with age.
  (b) Advantages: It can give more accurate estimates than simple random sampling when clear strata are present; gives separate estimates for each stratum. Disadvantages: Strata must be distinct, have means as different as possible, and have small variances.
5 **(a)** The population is divided into groups in terms of sex, age, social class, etc. and then a number of people (quota) for each group is decided. From then on, the choice of the sample units is left to the interviewer.
  (b) Quota sampling is used where numbers are large, where results are wanted quickly and cheaply e.g. national opinion polls.

Estimation and confidence intervals 3A

Key points to remember

1 Simple random sample
A simple random sample of size n consists of observations $X_1, X_2, \ldots, X_n$ from a population, where the X_i are independent random variables that have the same distribution as the population.

2 A statistic
If $X_1, X_2, \ldots, X_n$ is a random sample as defined above, then a statistic Y is a random variable consisting of any function of the X_i that involves no other unknown quantities.

3 Sampling distribution
The distribution of Y as defined above. This will be determined by the distribution of the X_i.

4 Estimator
A statistic that is used to estimate a population parameter. An estimate is a value of the estimator.

5 Unbiased estimator
If a statistic Y is used as an estimator for a population parameter θ and $E(Y) = \theta$, then Y is an unbiased estimator of θ.

6 Bias
If a statistic Y is used as an estimator for a population parameter θ and $E(Y) \neq \theta$, then Y is a biased estimator of θ. The bias is given by

$$\text{Bias} = E(Y) - \theta$$

Bias may be positive or negative.

12 Estimation and confidence intervals

7 Distribution of the sample mean $\overline{X}$
If you have a random sample of size n given by $X_1, X_2, \ldots, X_n$ from a population with mean μ and variance σ^2 then

$$E(\overline{X}) = \mu \text{ and } \operatorname{Var}(\overline{X}) = \frac{\sigma^2}{n}$$

In practice σ^2 is not usually known and then $\dfrac{s^2}{n}$ is used.

$$\text{If } X \sim N(\mu, \sigma^2) \text{ then } \overline{X} \sim N\!\left(\mu, \frac{\sigma^2}{n}\right)$$

Note: $\dfrac{\sigma}{\sqrt{n}}$ is known as the standard error of the mean.

8 Sample mean and sample variance
The sample mean $\bar{x}$ and the sample variance
$$s^2 = \frac{1}{n-1} \sum_{i=1}^{n} (x_i - \bar{x})^2$$
are unbiased estimates of the corresponding population parameters.

9 Confidence interval
An interval in which it is estimated, with a pre-selected degree of confidence, that the true value of the parameter being estimated will be contained.

This is not the same as the probability that the parameter lies in the interval.

10 Confidence limits for a normal mean
These are given by

$$\bar{x} \pm 1.96 \frac{\sigma}{\sqrt{n}} \text{ for a 95\% confidence interval}$$

$$\bar{x} \pm 1.6449 \frac{\sigma}{\sqrt{n}} \text{ for a 90\% confidence interval}$$

Example 1
Resistors produced on a machine have their resistance, in ohms, normally distributed with mean μ and standard deviation σ. A random sample of 6 resistors is taken and their resistances, $R_1, R_2, \ldots, R_6$, are measured.
Write down the distribution of Y when Y is defined as:

(a) $\displaystyle\sum_{i=1}^{6} R_i$

(b) $\dfrac{R_1 + 4R_6}{6}$

(c) Write down the distribution of $\overline{R}$.
(d) Use the above results to:
  (i) explain why $\overline{R}$ is an unbiased estimator of μ, but the estimator in part (b) is not an unbiased estimator of μ,
  (ii) find the bias of the estimator in (b).

Answer

(a) $E(Y) = E(R_1 + R_2 + \ldots + R_6)$
$ = E(R_1) + E(R_2) + \ldots + E(R_6)$
$ = \mu + \mu + \ldots + \mu$
$ = 6\mu$

$Var(Y) = Var(R_1 + R_2 + \ldots + R_6)$
$ = Var(R_1) + Var(R_2) + \ldots + Var(R_6)$
$ = \sigma^2 + \sigma^2 + \ldots + \sigma^2$
$ = 6\sigma^2$

$\sum_{i=1}^{6} R_i \sim N(6\mu, 6\sigma^2)$

Using **3**

(b) $E(Y) = E\left(\dfrac{R_1 + 4R_6}{6}\right)$
$ = \frac{1}{6}[E(R_1) + E(4R_6)]$
$ = \frac{1}{6}[E(R_1) + 4E(R_6)]$
$ = \frac{1}{6}[\mu + 4\mu]$
$ = \frac{5}{6}\mu$

$Var\left(\dfrac{R_1 + 4R_6}{6}\right) = \frac{1}{36}[Var(R_1) + Var(4R_6)]$
$\phantom{Var\left(\dfrac{R_1 + 4R_6}{6}\right)} = \frac{1}{36}[Var(R_1) + 16\,Var(R_6)]$
$\phantom{Var\left(\dfrac{R_1 + 4R_6}{6}\right)} = \frac{1}{36}[\sigma^2 + 16\sigma^2]$
$\phantom{Var\left(\dfrac{R_1 + 4R_6}{6}\right)} = \frac{17}{36}\sigma^2$

Using **3**

So: $\dfrac{R_1 + 4R_6}{6} \sim N(\frac{5}{6}\mu, \frac{17}{36}\sigma^2)$

(c) $\bar{R} \sim N(\mu, \frac{1}{6}\sigma^2)$

Using **7**

(d) (i) $E(\bar{R}) = \mu$ and so by definition it is unbiased.

Using **5**

$E\left(\dfrac{R_1 + 4R_6}{6}\right) = \frac{5}{6}\mu \neq \mu$ and so by definition is biased.

Using **6**

(ii) Bias $= \frac{5}{6}\mu - \mu = -\frac{1}{6}\mu$

Using **6**

Example 2

A random sample of 20 children is taken from a school. The frequency distribution of the number of children in the families of the children in the sample is shown below:

Number of children in family (x)	1	2	3	4	5	6	7	8
Frequency (f)	1	4	6	3	2	1	2	1

(a) Calculate an unbiased estimate for the mean number of children in a family.
(b) Calculate the standard error of the mean.

A further random sample of 20 children was selected and this sample had a mean of 3 children per family and variance 4.105.

14 Estimation and confidence intervals

(c) Treating the 40 results as a single sample, obtain further unbiased estimates of the mean and variance.

(d) Estimate the size of sample required to achieve a standard error of less than 0.35.

Answer

(a) By calculator $\sum fx = 77$ $\sum fx^2 = 367$ Using **7**

$$\bar{x} = \frac{77}{20} = 3.85$$

(b) $s_x^2 = \frac{1}{19}\left\{367 - \frac{(77)^2}{20}\right\}$ Using **8**

$= 3.713\,16 = 3.71$ (3 s.f.)

Standard error of the mean $= \dfrac{s}{\sqrt{n}}$ Using **8**

$$= \sqrt{\frac{3.713\,16}{20}}$$

$= 0.4309 = 0.431$ (3 s.f.)

(c) $\bar{y} = 3 \Rightarrow \sum fy = 20 \times 3 = 60$

$s_y^2 = 4.105 \Rightarrow \dfrac{\sum fy^2 - 20 \times 3^2}{19} = 4.105$

So $\sum fy^2 = 257.995$

Combined sample (w) of size 40 has

$$\sum fw = 77 + 60 = 137$$
$$\sum fw^2 = 367 + 257.995 = 624.995$$

The combined estimate for μ is $\bar{w} = \dfrac{137}{40} = 3.425$

and for σ^2 is $s_w^2 = \dfrac{624.995 - 40 \times 3.425^2}{39}$

$= 3.994$

(d) Using the best estimate for σ^2 (that based on the combined data) and **7**

the standard error of the mean $= \dfrac{s_w}{\sqrt{n}} = \sqrt{\dfrac{3.994}{n}}$

To achieve a standard error of < 0.35 you require

$$\sqrt{\frac{3.994}{n}} < 0.35$$
$$\sqrt{n} > \frac{\sqrt{3.994}}{0.35}$$
$$\sqrt{n} > 5.71$$
$$n > 32.6$$

So you need a sample of at least 33.

Worked examination question 1 [E]

Some ornithologists were studying a large group of plovers. A sample of 36 were measured and the length x centimetres of each plover was recorded. The results are summarised as follows:

$$\sum x = 561.6, \quad \sum x^2 = 8946.11$$

(a) Calculate unbiased estimates of the mean and the variance of the lengths of plovers.

Given that the standard deviation of the lengths of plovers is in fact 2.4 cm,
(b) find a 99% confidence interval for the mean length of plovers from this group.

Answer

(a) $\bar{x} = \dfrac{\sum x}{n} = \dfrac{561.6}{36} = 15.6$

$s^2 = \dfrac{1}{n-1}\left(\sum x^2 - n\bar{x}^2\right)$ Using **8**

$= \dfrac{1}{35}(8946.11 - 36 \times 15.6^2) = 5.29$

(b) The 99% confidence limits are   Using **10**

$$\bar{x} \pm 2.5758 \dfrac{\sigma}{\sqrt{n}} = 15.6 \pm 2.5758 \times \dfrac{2.4}{\sqrt{36}} = 15.6 \pm 1.03$$

The 99% confidence interval is (14.6, 16.6).

Worked examination question 2 [E]

A quality control inspector is about to take a random sample of 5 metal bars from a production line and calculate 95% confidence limits for the mean length of bars from this production line.
(a) State the probability that these limits will contain the mean of the population.
The sample had lengths, in mm, as shown:

15.63, 15.70, 15.61, 15.67, 15.68.

(b) Find an unbiased estimate for the variance of the length of the bars.
The inspector consulted past records and discovered that over a long period the standard deviation had been 0.041.
(c) Using this value for the standard deviation of the population, determine the 95% confidence limits for the mean.
(d) State an additional assumption you need to make for these limits to be valid.
The probability that the mean of the population lies between these limits is not the same as the answer to part (a).
(e) Explain why this is so.

Answer

(a) The probability that the limits will contain the mean of the population is 0.95. *Using 9*

(b) $s^2 = \dfrac{1}{n-1}\left\{\sum x^2 - \dfrac{(\sum x)^2}{n}\right\}$ *Using 8*

$= \dfrac{1}{4}\left\{1225.8703 - \dfrac{78.29^2}{5}\right\}$

$= 0.00137$

(c) The 95% confidence limits are   *Using 10*

$$\bar{x} \pm 1.96 \dfrac{\sigma}{\sqrt{n}} = \dfrac{78.29}{5} \pm 1.96 \times \dfrac{0.041}{\sqrt{5}}$$

$= 15.658 \pm 0.035938$

The 95% confidence interval is (15.62, 15.69).

(d) The lengths of the rods are, or can be considered to be, normally distributed.

(e) The probability in the first case is the probability of the interval being one that contains the mean. This is not the same as the probability of the mean being within the limits of the interval. The population mean is either in or not in the interval with probability 1 or 0.   *Using 9*

Revision exercise 3A

1. A simple random sample $X_1, X_2, \ldots, X_6$ is taken from a population distributed $N(18, 3.2^2)$.
   (a) Write down the distribution of X_3.
   (b) Write down the distribution of $X_1 + X_3 + X_6$.
   (c) Write down the distribution of $\overline{X}$.

2. A random sample of size 36 was taken from a population distributed $N(\mu, 3.9^2)$. The value of the sample mean $\bar{x}$ was 15.6.
   (a) Find a 90% confidence interval for μ.
   It is believed that the value of μ is 17.0.
   (b) Use your confidence interval to comment on this belief. [E]

3 The random variable X is normally distributed with mean μ and variance σ^2
(a) Write down the distribution of the sample mean $\overline{X}$ based on samples of size n.
(b) From time to time a company manufacturing plastic windows needs to check the lengths of a certain component to ensure that no change has occurred. It is known from experience that the standard deviation of the length is 0.23 mm. The company wishes to take a sample of size n and to calculate a 95% confidence interval for the mean length of the population. The width of the confidence interval must be no more than 0.05 mm. Calculate a minimum value for n.

Test yourself

What to review

If your answer is incorrect:

1 A simple random sample $X_1, X_2, \ldots, X_{10}$ is taken from a population with mean μ and variance σ^2.
(a) Write down the distribution of X_i.
(b) Write down the distribution of $X_1 + X_5$.

Review Heinemann Book S3 pages 17–18

2 A simple random sample $X_1, X_2, \ldots, X_5$ is taken from a population distributed $N(0, 2.4^2)$.
Write down the distribution of $\overline{X}$.

Review Heinemann Book S3 pages 22–23

3 (a) Explain what is meant by a 90% confidence interval for a population mean.
The lengths of components produced by an automatic lathe are known to be normally distributed with standard deviation 0.03 mm. A sample of 15 components is found to have a mean length of 18.32 mm.
(b) Estimate the standard error of the mean. Use this to estimate a 90% confidence interval for the mean length of the population.
(c) Find the minimum size of the sample of components which must be taken if the width of the confidence interval is to be at most 0.02 mm.

Review Heinemann Book S3 pages 23–26 and 36–43

Test yourself answers

1 (a) $N(\mu, \sigma^2)$ (b) $N(2\mu, 2\sigma^2)$ 2 $N(0, 1.152)$ 3 (b) 0.007 75, (18.307, 18.333) (c) 25

Hypothesis tests for means

3B

Key points to remember

1 Central limit theorem

If $X_1, X_2, \ldots, X_n$ is a random sample of size n from a population, with mean μ and variance σ^2 then

$$\overline{X} \sim N(\mu, \frac{\sigma^2}{n}) \text{ as } n \to \infty.$$

2 Single-sample problems

If a random sample $X_1, X_2, \ldots, X_n$ is selected from $N(\mu, \sigma^2)$ then

$$\frac{\overline{X} - \mu}{\sigma/\sqrt{n}} \sim N(0, 1^2)$$

The central limit theorem implies that for large samples

$$\frac{\overline{X} - \mu}{S/\sqrt{n}} \text{ is approximately } \sim N(0, 1^2) \qquad S^2 = \frac{1}{n-1}\sum(X - \overline{X})^2$$

Since for large n, S is a close approximation to σ.

3 Two-sample problems

If random samples of n_x X's from $N(\mu_x, \sigma_x^2)$ and n_y Y's from $N(\mu_y, \sigma_y^2)$ are selected then under suitable conditions

$$\frac{(\overline{X} - \overline{Y}) - (\mu_x - \mu_y)}{\sqrt{\left(\frac{\sigma_x^2}{n_x} + \frac{\sigma_y^2}{n_y}\right)}} \sim N(0, 1^2)$$

If $\mu_x = \mu_y$ these reduce to

$$\frac{(\overline{X} - \overline{Y})}{\sqrt{\left(\frac{\sigma_x^2}{n_x} + \frac{\sigma_y^2}{n_y}\right)}} \sim N(0, 1^2)$$

and

$$\frac{(\overline{X} - \overline{Y})}{\sqrt{\left(\frac{S_x^2}{n_x} + \frac{S_y^2}{n_y}\right)}} \sim N(0, 1^2)$$

The central limit theorem implies that for large samples

$$\frac{(\overline{X} - \overline{Y}) - (\mu_x - \mu_y)}{\sqrt{\left(\frac{S_x^2}{n_x} + \frac{S_y^2}{n_y}\right)}} \text{ is approximately } \sim N(0, 1^2)$$

Since for large n, S is a close approximation to σ.

Hypothesis tests for means

Example 1
A tow rope is intended to have a breaking strength of 10 000 N. It is known that, due to imperfections in the rope, the breaking strength is normally distributed with a standard deviation of 150 N. A sample of 10 ropes gave a mean breaking strength of 9500 N. Is there evidence at the 5% significance level of a decrease in breaking strength?

Answer
Since you are dealing with a normal distribution with known variance you can use **2**.
$H_0: \mu = 10\,000, \quad H_1: \mu < 10\,000$

Test statistic $z = \dfrac{\bar{x} - \mu}{\sigma/\sqrt{n}} = \dfrac{9500 - 10\,000}{\dfrac{150}{\sqrt{10}}} = -10.54$

Critical region from Table 4 in Book S3 is $Z \leqslant -1.6449$.
The value is within the critical region. There is evidence to suggest that the mean has decreased.

Example 2
The weights of French loaves, X kg, produced by a baker are normally distributed. It is required that the mean weight of the loaves should be 0.5 kg. A sample of 50 loaves yielded the following summary statistics.
$$\sum x = 24, \quad \sum x^2 = 11.955$$
Is there any indication, at the 5% level of significance, that the mean mass of the loaves is not 0.5 kg?

Answer
Since the distribution is normal, and the number in the sample is large, you can use **2**.
$H_0: \mu = 0.5, \quad H_1: \mu \neq 0.5$

$$\bar{x} = \frac{24}{50} = 0.48$$

$$s^2 = \frac{1}{n-1}\left(\sum x^2 - n\bar{x}^2\right)$$
$$= \frac{1}{49}(11.955 - 50 \times 0.48^2)$$
$$= 0.008\,877\,55$$
$$s = 0.094\,220\,7$$

Test statistic $z = \dfrac{\bar{x} - \mu}{s/\sqrt{n}}$

$= \dfrac{0.48 - 0.5}{\dfrac{0.094\,220\,7}{\sqrt{50}}}$

$= -1.50$ (3 s.f.)

> Since σ is unknown and the sample is large you use s.

Critical region from Table 4 in Book S3 is $Z < -1.96$ and $Z > 1.96$.
Since -1.50 is not in the critical region, there is insufficient evidence to suggest that the mean weight has changed.

Example 3
It is claimed that on a standard test given in all schools at age 11, boys in single sex schools score higher marks than those from mixed schools. The scores, x, of a random sample of 120 boys from single sex schools and the scores, y, of a random sample of 130 boys from mixed schools were recorded. The results were as follows

$$\sum x = 13\,800, \ \sum y = 14\,300$$

The variance for the total population of boys from both single sex and mixed schools was 400.
Test, at the 5% significance level, whether or not the results support the claim.

Answer

$H_0: \mu_x = \mu_y, \quad H_1: \mu_x > \mu_y$

Test statistic $z = \dfrac{(\bar{x} - \bar{y}) - (\mu_x - \mu_y)}{\sqrt{\left(\dfrac{\sigma_x^2}{n_x} + \dfrac{\sigma_y^2}{n_y}\right)}}$

Using **3**

$= \dfrac{(115 - 110) - 0}{\sqrt{400}\sqrt{\left(\dfrac{1}{120} + \dfrac{1}{130}\right)}}$

Note $\sigma_x^2 = \sigma_y^2$, and $\mu_x = \mu_y$ is assumed under H_0.

$= 1.97$

Critical region from Table 4 in Book S3 is $Z \geqslant 1.6449$
Since 1.97 is in the critical region, there is evidence to support the claim.

Example 4
Beth and Tara live next door to each other and attend the same school. Each day they leave home at the same time but travel to school by different routes. Tara claims that Beth gets to school on average 5 minutes after she does. Both girls recorded their journey times, to the nearest minute, over 5 weeks. The results are summarised below:

	Mean	Variance	Sample size
Beth	25	9.73	25
Tara	21	6.85	25

Assuming that the times are normally distributed, test at the 5% significance level whether or not Tara's claim is justified.

Answer

Let X represent Beth's times and Y represent Tara's times.

$H_0: \mu_x = \mu_y + 5$, $\quad H_1: \mu_x \neq \mu_y + 5$ Using **3**

Test statistic $z = \dfrac{(\bar{x} - \bar{y}) - (\mu_x - \mu_y)}{\sqrt{\left(\dfrac{s_x^2}{n_x} + \dfrac{s_y^2}{n_y}\right)}}$

$= \dfrac{(25 - 21) - 5}{\sqrt{\left(\dfrac{9.73}{25} + \dfrac{6.85}{25}\right)}}$

$= -1.23$ (3 s.f.)

The critical region is $Z < -1.9600$ or $Z > 1.9600$.
Since -1.23 is not in the critical region, there is evidence to suggest that the claim is justified.

Worked examination question 1 [E]

The owners of a Chinese take-away kept records for many years of their daily takings and were concerned that their mid-week takings were lower than those for the rest of the week.

The takings for Wednesdays followed the normal distribution with mean £130.00 and standard deviation £27.60. The owners tried to boost their mid-week takings by having special offers available to customers on Wednesdays for a period of 30 weeks. The mean takings for these 30 Wednesdays was £145.67.

Treating these 30 Wednesdays as a random sample, the owners decided to test whether or not there was evidence that the special offers increased takings on Wednesdays.

(a) Stating your hypotheses clearly, carry out this test using a 1% level of significance.

(b) State an assumption you have made about the effect of the special offers on the standard deviation of the takings on Wednesdays.

Answer

(a) $H_0: \mu = 130, \quad H_1: \mu > 130$

$z = \dfrac{\bar{x} - \mu}{\dfrac{\sigma}{\sqrt{n}}} = \dfrac{145.67 - 130}{\dfrac{27.60}{\sqrt{30}}} = 3.11$

1% critical value is 2.3263
The result is significant. Reject H_0 – there is evidence to suggest that special offers have boosted sales.

(b) The standard deviation of the takings on Wednesdays was not affected by the special offers.

Worked examination question 2 [E]

(a) State briefly the Central Limit Theorem.
A local government officer is monitoring equality of opportunity between disabled and non-disabled employees of the authority. She

22 Hypothesis tests for means

has chosen two random samples, one of each type of employee. For each example, she has recorded the numbers of employees in each of 5 salary ranges. The data are recorded in the table.

Salary (£1000s)	0–	10–	20–	30–	40–
Mid-value, x	5	15	25	35	45
Disabled, f_d	22	24	7	5	2
Non-disabled, f_n	39	45	33	22	11

You may assume that

$$\sum f_d = 60, \quad \sum f_d x = 910, \quad \sum f_d x^2 = 20\,500,$$
$$\sum f_n = 150, \quad \sum f_n x = 2960, \quad \sum f_n x^2 = 80\,950$$

The officer wishes to test whether disabled employees are paid less than non-disabled employees by this authority. She decides to use the Central Limit Theorem for both samples.

(b) State the null and alternative hypotheses.

(c) Carry out the test at the 5% level of significance.

Figure 1 shows a histogram for the salaries of the non-disabled employees.

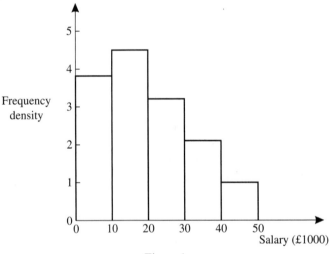

Figure 1

(d) By considering the histogram, give a reason why it is necessary to use the Central Limit Theorem for the non-disabled sample, rather than assuming that it was drawn from a normal distribution.

Answer

(a) If $X_1, X_2, \ldots, X_n$ is a random sample of size n from a population, not necessarily normal, with mean μ and variance σ^2 then as $n \to \infty$ $\overline{X} \sim N\left(\mu, \dfrac{\sigma^2}{n}\right)$.

(b) $H_0: \mu_D = \mu_N$, $H_1: \mu_D < \mu_N$
(c) $\bar{x}_D - \bar{x}_N = 15.16 - 19.73 = -4.56$

$$\sigma_D^2 = \frac{1}{59}(20\,500 - 60 \times 15.1\dot{6}^2) = 113.531$$

$$\sigma_N^2 = \frac{1}{149}(80\,950 - 150 \times 19.7\dot{3}^2) = 151.2707$$

Test statistic $z = \dfrac{\bar{x}_D - \bar{x}_N}{\sqrt{\left(\dfrac{\sigma_D^2}{60} + \dfrac{\sigma_N^2}{150}\right)}}$

$= \dfrac{-4.5\dot{6}}{\sqrt{\left(\dfrac{113.531}{60} + \dfrac{151.2707}{150}\right)}}$

$= -2.6813$

The 5% 1-tail critical value is -1.6449.
The result is significant. There is evidence to suggest that the disabled are paid less.
(d) The histogram is skewed so the distribution is not normal – you need the central limit theorem to get $\bar{X} \sim$ Normal.

Revision exercise 3B

1 In an examination, the marks of all the candidates were distributed with a mean of 55 marks and a standard deviation of 12 marks. The Director of Education for a certain region found that for a random sample of 100 pupils in that region, the mean mark was 60.
(a) Test at the 1% significance level whether or not there is evidence to support the Director's claim that pupils from his region had a higher mean mark than that of all candidates.
(b) Explain how the central limit theorem has been used in this example.

2 It is thought that a couple's courtship was longer before a second marriage than before a first marriage. In a survey into courtship a sociologist randomly sampled 100 couples that were on their first marriage and recorded the lengths of time, x months, of their courtship. The sociologist also recorded the lengths of time y months, of 90 couples on their second marriage. The results for the two groups were as follows:

$$\sum x = 26\,500, \quad \sum y = 24\,400$$

The best estimate of the standard deviation, s, when the two groups were combined was 45.
Test at the 5% significance level whether or not courtship is longer before a second marriage.

3. A sociologist was studying the smoking habits of adults. A random sample of 300 adult smokers from a low-income group and an independent random sample of 400 adult smokers from a high-income group were asked their weekly expenditure on tobacco. The results were as follows.

	n	mean	s.d.
Low-income group	300	£6.40	£6.69
High-income group	400	£7.42	£8.13

(a) Using a 5% significance level, test whether or not the two groups differ in the mean amount spent on tobacco.
(b) Explain briefly the importance of the central limit theorem in this example. [E]

4. From past evidence the number of errors made by rats in a maze-running experiment is known to be normally distributed with a mean of 14.05 and a standard deviation of 11.53.
In an experiment into the effects of a particular drug on rats' maze-learning ability, a randomly selected group of 20 rats were given the drug and the mean number of errors noted was 15.
Test at the 5% level of significance whether or not the drug increased the number of errors made by the rats.

Test yourself

What to review

If your answer is incorrect:

1. Jam is packed into tins of advertised weight 1 kg. The weight, X, of a randomly selected tin is normally distributed about a target weight with a standard deviation of 12 grams. In order to ensure that the tins do not have weights less than 1 kg the target weight is set at 1.03 kg.
A random sample of 20 tins is taken and the total weight of the tins $\sum x = 20.4$ kg.
Test at the 1% significance level whether or not the mean weight of the tins is less than the target weight.

Review Heinemann Book S3 pages 43–50

2 In a survey of 100 school children, the children were found to have a mean weight of 39 kg, with a standard deviation of 3.5 kg. In a previous survey of 80 school children the mean weight was 37.5 kg and the standard deviation was 3.8 kg.
(a) Test at the 5% level of significance to see whether or not there is evidence that the mean weight of children has increased.
(b) State the central limit theorem and say why it can be used in this example.

Review Heinemann Book S3 pages 43–50

3 In a study of train services into a London terminal, the trains of operator A are compared for lateness of arrival with those of operator B. For a random sample of 35 trains run by operator A the number of minutes late, x, was recorded for each train. Similarly, for operator B, the number of minutes late, y, was recorded for each of 35 trains. The results are summarised below.

$$\sum x = 26.25$$
$$\sum y = 20.4$$

(a) Assuming that the common standard deviation is 2.205 min, test at the 5% level of significance whether or not there is any difference between the mean number of minutes late for the two operators.
(b) State any assumption you have made in order to do this test.

Review Heinemann Book S3 pages 50–53

4 A machine fills packets of sugar. Each day a random sample of 50 packets is taken and weighed.
After the machine had broken down and been repaired, it was decided to take a random sample of 60 packets and compare the mean weight of the sample to the mean weight of the sample taken on the last day before the machine broke down. The weight, y kg, of each of the 60 packets was recorded and the weights, x kg, of each of the packets in the previous sample found.
The data from the two samples are summarised below:

| Before breakdown | $\sum x = 50.5$ | $\sum x^2 = 52.279$ |
| After breakdown | $\sum y = 63.0$ | $\sum y^2 = 67.979$ |

(a) Test at the 5% level of significance whether or not the mean weight of a packet has changed.
(b) Explain how the central limit theorem has been used in this problem.

Review Heinemann Book S3 pages 50–59

Test yourself answers

1 $H_0: \mu = 1.03$, $H_1: \mu < 1.03$. Test stat $= -3.73$. Critical value $= -2.3263$. Reject H_0 – evidence that the mean is less than the target weight.

2 (a) $H_0: \mu_A = \mu_B$, $H_1: \mu_A > \mu_B$. Critical value $= 1.6449$. Test stat $= 2.73$. Reject H_0 – evidence that the mean weight has increased.
   (b) For large samples $\bar{X} \sim N\left(\mu, \dfrac{\sigma^2}{n}\right)$. In this case n is large (100 or 80).

3 (a) $H_0: \mu_A = \mu_B$, $H_1: \mu_A \neq \mu_B$. Critical values $= \pm 1.9600$. Test stat $= 0.317$. No evidence of a difference in the mean number of minutes late for the two operators.
   (b) Normality of the distributions.

4 (a) $H_0: \mu_x = \mu_y$, $H_1: \mu_x \neq \mu_y$. $\mu_x = 1.01$, $s_x^2 = 0.026$, $\mu_y = 1.05$, $s_y^2 = 0.031$. Test stat $= -1.2423$. Critical values ± 1.9600. Reject H_0 – no evidence of a difference between the mean weight before and after repairs.
   (b) Sample size is large so test statistic is $\sim N(0, 1)$.

Goodness of fit

4A

Key points to remember

1. Observed and expected frequencies
O_i are the observed frequencies.
E_i are the frequencies expected if the proposed model is applied.

2. Hypotheses
Null hypothesis H_0: usually specified in terms of a model for goodness of fit.
e.g. H_0: A binomial distribution is a suitable model.
Alternative hypothesis H_1: usually specified in terms of a model for goodness of fit.
e.g. H_1: A binomial distribution is not a suitable model.

3. Test statistic

$$\sum \frac{(O_i - E_i)^2}{E_i}$$

4. Chi-squared
Provided none of the $E_i < 5$

$$\sum \frac{(O_i - E_i)^2}{E_i} \approx \chi^2_\nu$$

where ν is as defined in **6** below.
If any $E_i < 5$ then cells are combined.

5. Constraints
Features that affect the evaluation of expected frequencies
e.g. calculation of a mean.

6. Degrees of freedom ν
Number of cells after combining – number of constraints.

7 Significance level
The probability of rejecting the null hypothesis when the null hypothesis is true.

8 Critical value
The value of χ^2_ν that is exceeded with probability equal to the significance level. The critical value can be found in the χ^2 table on page 166 of Book S3.

9 Possible models
If n = number of expected values after combining:

	Degrees of freedom	
Distribution	Parameters known	Parameters not known
Discrete uniform	$n-1$	
Binomial	$n-1$	$n-2$
Poisson	$n-1$	$n-2$
Normal	$n-1$	$n-3$ both unknown $n-2$ one unknown
Continuous uniform	$n-1$	

Example 1

A six-sided die being used at a gambling centre seemed to be favouring certain numbers. Over 120 throws the following results were recorded.

Number on die	1	2	3	4	5	6
Frequency	18	35	12	27	18	10

Test at the 1% significance level whether or not these data support the claim that the die is biased.

Answer

If the die is not biased you would expect a discrete uniform distribution to be an appropriate model for the results.
H_0: A discrete uniform distribution is a suitable model for these results.
H_1: A discrete uniform distribution is not a suitable model for these results.

28 Goodness of fit

The expected frequencies will be given by $\dfrac{\sum \text{frequencies}}{n} = \dfrac{120}{6} = 20$

Number on die	1	2	3	4	5	6	Total
O_i	18	35	12	27	18	10	120
E_i	20	20	20	20	20	20	120
$\dfrac{(O_i - E_i)^2}{E_i}$	0.2	11.25	3.2	2.45	0.2	5	22.3

Note: $\sum E_i$ must be equal to $\sum O_i$.

$$\nu = n - 1 = 6 - 1 = 5$$

Using **6** and **9**

χ_5^2 (0.01) critical value = 15.086

Test statistic $\sum \dfrac{(O_i - E_i)^2}{E_i} = 22.3 > 15.086$

Using **3**

There is sufficient evidence to reject H_0. A discrete uniform distribution is not a good model. There is evidence to suggest the die is biased.

Example 2
On 30 consecutive weekdays the numbers of pieces of junk mail x received by a teacher through the post were as follows:.

4, 2, 1, 0, 7, 3, 2, 2, 1, 0, 1, 5, 2, 3, 0,
0, 1, 2, 2, 4, 3, 3, 1, 5, 4, 4, 4, 6, 1, 2.

(a) Find values for a, b and c in the following frequency table.

x	0	1	2	3	4	5	6	7
Observed frequency	4	a	b	4	5	c	1	1

(b) It is thought that a Poisson distribution might be a suitable model for these data. Complete the following table by finding the values of r, s and t.

x	0	1	2	3	4	5	6	≥ 7
Expected frequency	2.463	r	7.695	s	4.008	2.004	0.835	t

(c) Stating your hypotheses clearly, test at the 5% level of significance whether or not a Poisson distribution is a suitable model for these data.

Answer
(a) By observation it can be seen that $a = 6$, $b = 7$, $c = 2$.

(b) $\lambda = \dfrac{\sum xf_x}{N} = \dfrac{75}{30} = 2.5$

$r = \dfrac{e^{-\lambda}\lambda^1}{1!} \times 30 = 6.156$, $\quad s = 7.695 \times \dfrac{\lambda}{3} = 6.413$,

$t = 30 - (2.463 + r + 7.695 + s + 4.008 + 2.004 + 0.835) = 0.426$

(c) H_0: A Poisson distribution is a suitable model.
H_1: A Poisson distribution is not a suitable model.

Combining expected values < 5

x	≤ 1	2	3	≥ 4	Totals
O_x	10	7	4	9	30
E_x	8.619	7.695	6.413	7.273	30
$\dfrac{(O_x - E_x)^2}{E_x}$	0.221	0.063	0.908	0.410	1.602

You have estimated λ so $\nu = 4 - 1 - 1 = 2$
χ_2^2 (5%) critical value $= 5.991$

Using **6** and **9**

Test statistic $= \sum \dfrac{(O_x - E_x)^2}{E_x} = 1.602 < 5.991$.

Using **3**

There is insufficient evidence to reject H_0. The Poisson distribution is a suitable model.

Example 3

80 Petri dishes containing cotton wool each have 10 seeds sown in them. Germination of the seeds is as shown in the table.

Number of seeds germinating	0	1	2	3	4	5	6	7	8	9	10
Observed frequency	4	13	30	17	8	5	2	1	0	0	0

(a) Give reasons why a binomial distribution might be a suitable model for these data.

(b) It is suggested that a B(10, 0.25) distribution would be a good model. Find values for p, q and r in the table of expected frequencies given below.

Number of seeds germinating	0	1	2	3	4	5	6	7	8	9	10
Expected frequency	4.504	p	22.528	20.024	q	4.672	r	0.248	0.032	0	0

(c) Using a significance level of 1% test this model for suitability.

Answer

(a) Germination has two outcomes – success or failure.
There are a fixed number of trials.
The trials are independent.
The probability of success is constant.

(b) The binomial cumulative distribution table on page 159 in Book S3 can be used to find p, q and r.

$p = (0.2440 - 0.0563) \times 80 = 15.016$
$q = (0.9219 - 0.7759) \times 80 = 11.680$
$r = (0.9965 - 0.9803) \times 80 = 1.296$

The values in the table above and these answers for p, q and r were found using the tables in the booklet provided in the examination.
If a calculator is used minor differences in some values occur. Either method is acceptable in the examination.

(c) H_0: B(10, 0.25) is a suitable model.
H_1: B(10, 0.25) is not a suitable model.
Combining expected frequencies < 5.

i	≤ 1	2	3	4	≥ 5	Total
O_i	17	30	17	8	8	80
E_i	19.52	22.528	20.024	11.680	6.248	80
$\dfrac{(O_i - E_i)^2}{E_i}$	0.325	2.478	0.457	1.159	0.491	4.91

You have not estimated p so $\nu = 5 - 1 = 4$
χ^2_4 (1%) critical value $= 13.277$

Test statistic $\sum \dfrac{(O_i - E_i)^2}{E_i} = 4.91 < 13.277$

There is insufficient evidence to reject H_0. B(10, 0.25) is a suitable model.

Worked examination question 1 [E]

A biologist was attempting to test a theory that a species of leaf insect has a lifespan whose distribution can be modelled as uniform between 0 and 20.5 days. He collected the data shown in the table.

Lifespan (in days, to the nearest whole day)	0–2	3–5	6–10	11–20
Number of insects	38	53	75	112

He performed a goodness of fit test at the 5% level of significance and discovered that the test statistic was 9.098. He did not need to combine any classes.

(a) State, with a justification, the conclusion he should have drawn.

He observed that, in the sample he had taken, no insect survived for longer than 16 days (to the nearest whole day). He decided to revise his theory by reducing the upper limit of the distribution to 16.5.

(b) Using the new model, perform a goodness of fit test at the 5% significance level, and comment on whether this refinement seems justified.

Answer

(a) For a continuous uniform distribution model
$$\nu = n - 1 = 4 - 1 = 3.$$
From the table χ^2_3 (5%) critical value $= 7.815$
The test statistic $= 9.098 > 7.815$
This is a significant result so there is evidence to reject H_0 and U[0, 20.5] is not a suitable model.

(b) H_0: U[0, 16.5] is a suitable model
H_1: U[0, 16.5] is not suitable model

The new expected values are given by $\dfrac{\sum O_i}{16.5} \times$ class width.

Lifespan	0 0–2 2.5	3–5 5.5	6–10 10.5	11–16 16.5	Totals
O_1	38	53	75	112	278
E_1	42.12	50.55	84.24	101.09	278
$\dfrac{(O_i - E_i)^2}{E_i}$	0.403	0.119	1.014	1.177	2.713

As before $\nu = 3$ and χ_3^2 (5%) critical value $= 7.815$

Test statistic $\sum \dfrac{(O_i - E_i)^2}{E_i} = 2.713 < 7.815$ 　　　　Using **3**

There is insufficient evidence to reject H_0. The continuous uniform model $U[0, 16.5]$ is a suitable model, so the refinement seems justified.

Worked examination question 2 [E]
A hamster breeder is studying the weight of adult hamsters. Each hamster from a random sample of 50 hamsters is weighed and the results, given to the nearest g, are recorded in the following table.

Weight (g)	85–94	95–99	100–104	105–109	110–119
Frequency	6	9	17	14	4

(a) Show that an estimate of the mean weight of the hamsters is 102 g.

The breeder proposes that the weight of an adult hamster, in g, should be modelled by the random variable W having a normal distribution with standard deviation 6. The breeder fits a normal distribution and obtains the following expected frequencies.

W	$W \leqslant 94.5$	$94.5 < W \leqslant 99.5$	$99.5 < W \leqslant 104.5$	$104.5 < W \leqslant 109.5$	$W > 109.5$
Expected frequency	r	11.64	s	11.64	t

(b) Find the values of r, s and t.
(c) Stating your hypotheses clearly, test at the 10% level of significance whether or not a normal distribution with a standard deviation of 6 is a suitable model for W.
(d) Calculate a 95% confidence interval for the mean weight of adult hamsters.
(e) Explain the importance of the test in part (c) to the calculation in part (d).

Answer
(a) $\bar{x} = \dfrac{6 \times 89.5 + 9 \times 97 + 17 \times 102 + 14 \times 107 + 4 \times 114.5}{50}$

$= \dfrac{5100}{50}$

$= 102$

(b) $r = 50 \times P\left(Z < \dfrac{94.5 - 102}{6}\right)$
$= 50 \times P(Z < -1.25)$
$= 5.28$

$t = r$ by symmetry $= 5.28$

$s = 50 - (5.28 + 11.64 + 11.64 + 5.28)$
$= 50 - 33.84$
$= 16.16$

(c) H_0: $N(\mu, 6^2)$ is a good model.
H_1: $N(\mu, 6^2)$ is not a good model.

W	$W \leq 94.5$	$94.5 < W \leq 99.5$	$99.5 < W \leq 104.5$	$104.5 < W \leq 109.5$	$W > 109.5$
O_i	6	9	17	14	4
E_i	5.28	11.64	16.16	11.64	5.28
$\dfrac{(O_i - E_i)^2}{E_i}$	0.0982	0.5988	0.0437	0.4785	0.3103

Test statistic $\sum \dfrac{(O_i - E_i)^2}{E_i} = 1.5295$

Using **3**

μ is estimated from the observations and there is the usual constraint that totals have to agree so $\nu = 5 - 1 - 1 = 3$.

Using **6** and **9**

From the table χ_3^2 (10%) critical value $= 6.251$
Test statistic $= 1.5295 < 6.251$ so the value is not significant.
$N(\mu, 6^2)$ is a good model in this case.

(d) 95% confidence interval is
$\bar{x} \pm 1.96 \dfrac{\sigma}{\sqrt{n}} = 102 \pm 1.96 \dfrac{6}{\sqrt{50}} = (100, 104)$

(e) The test in (c) confirms the normal distribution that is necessary for (d) to apply.

Revision exercise 4A

1 A square spinner for use in a child's game has the four numbers 1, 3, 4 and 6 on its edges. The manufacturer claims that the probability that the spinner lands on each of the four numbers is $\frac{1}{4}$. Before a box of spinners leaves the factory a randomly chosen spinner is tested by spinning it 20 times and recording the number of sixes.

(a) Using a 5% significance level, find the critical regions for a two-tailed test of the hypothesis that the probability of the spinner landing on a six is $\frac{1}{4}$.

(b) Explain how a spinner could pass this test but still not satisfy the manufacturer's claim.

(c) Suggest an alternative test that the manufacturer could use to refine the testing procedure.

A quality control inspector tests a spinner by spinning it 40 times and recording the outcomes. His results are given below.

Outcome	1	3	4	6
Frequency	7	6	16	11

(d) Test the manufacturer's claim at the 5% significance level, stating your null and alternative hypotheses clearly.

(e) Give one advantage and one disadvantage of the test used in (d) compared with that used by the manufacturer in (a). [E]

2 A random sample of 300 football matches was taken and the number of goals scored in each match was recorded. The results are given in the table below.

Number of goals	0	1	2	3	4	5	6	7
Frequency	33	55	80	56	56	11	5	4

(a) Show that an unbiased estimate of the mean number of goals scored in a football match is 2.4 and find an unbiased estimate of the variance.

It is thought that a Poisson distribution might provide a good model for the number of goals scored per match.

(b) State briefly an implication of a Poisson model on goal scoring at football matches.

Using a Poisson distribution, with mean 2.4, expected frequencies were calculated as follows:

Number of goals	0	1	2	3	4	5	6	7
Expected frequency	s	65.3	t	62.7	37.6	18.1	7.2	2.5

(c) Find the values of s and t.

(d) State clearly the hypotheses required to test whether or not a Poisson distribution provides a suitable model for these data.

In order to carry out this test the class for 7 goals is redefined as 7 or more goals.

(e) Find the expected frequency for this class.

The test statistic for the test in part (d) is 15.6 and the number of degrees of freedom used is 5.

(f) Explain fully why there are 5 degrees of freedom.

(g) Stating clearly the critical value used, carry out the test in part (d), using a 5% level of significance. [E]

3 A biologist is trying to model the number of female kittens born in a litter. She has examined 250 litters of size 5 and obtained the following results.

Number of females	0	1	2	3	4	5
Number of litters	2	40	90	85	30	3

(a) Test at the 5% level of significance whether or not the binomial distribution with parameters $n = 5$ and $p = 0.5$ is an adequate model for these data.

The biologist decides to refine the model and uses the above data to estimate the proportion p of kittens that are female.

(b) Find this estimate of p.

Using this value of p she went on to calculate expected frequencies and the statistic $\sum \dfrac{(O-E)^2}{E}$ and obtained the value 10.951 (no classes were combined).

(c) Stating your hypotheses clearly, explain what conclusions the biologist should deduce from this value. Use a 5% level of significance.

(d) What are the implications of this analysis for the number of female kittens in a litter of 5?

4 During an experiment involving the weight of eggs from which ducklings failed to hatch, a laboratory assistant wondered whether or not the weights of the eggs could be modelled by a normal distribution $N(68, 5^2)$. The weights of 1000 such eggs were recorded with the following results:

Weight (g)	$\leqslant 62$	>62–64	>64–67	>67–70	>70–73	>73–76	>76
Frequency	81	187	204	208	160	100	60

Using a 5% significance level, test whether or not $N(68, 5^2)$ is a suitable model.

Test yourself	What to review
	If your answer is incorrect:

1 As part of a goodness of fit test on a uniform distribution the following expected and observed frequencies were recorded:

Observed frequency	20	16	25	18	21
Expected frequency	20	20	20	20	20

(a) Find for these data the value of the test statistic
$$\sum \frac{(O - E)^2}{E}.$$
(b) Assuming that there are 4 degrees of freedom, find the critical value at the 5% significance level.
(c) What conclusion would you draw from parts (a) and (b)?

Review Heinemann Book S3 pages 79–81

2 The binomial distribution B(5, 0.5) is to be used as a model. Assuming that the sum of the observed frequencies = 100, calculate the expected frequencies.

Review Heinemann Book S3 pages 81–85

3 The Poisson distribution Po(1.7) is to be used as a model. Assuming that the sum of the observed frequencies = 50, calculate the expected frequencies of the first three cells. Calculate the total of the expected frequencies for the cells that remain.

Review Heinemann Book S3 pages 86–88

4 A sample of 50 peaches is taken from baskets at the supermarket. The peaches have the following distribution of diameters:

Diameter (cm)	<6	6–	7–	8–	≥9
Frequency	4	13	19	8	6

It is suggested that these diameters can be modelled by a N(7, 0.9^2) distribution.
Calculate the expected frequencies.

Review Heinemann Book S3 pages 90–98

Test yourself answers

1 (a) 2.3 (b) 9.488 (c) A uniform distribution is a good model.
2 3.125, 15.625, 31.25, 31.25, 15.625, 3.125
3 9.1342, 15.5281, 13.1989, 12.1388
4 6.675, 18.325, 18.325, 6.015, 0.66

Contingency tables 4B

Key points to remember

1 **Contingency table**
A table representing the results of a sample, where each member of the sample can be classified according to two criteria.

		Criterion 1		
		Level 1	Level 2	Level 3
Criterion 2	Level 1	Frequency cell (1, 1)	Frequency cell (1, 2)	Frequency cell (1, 3)
	Level 2	Frequency cell (2, 1)	Frequency cell (2, 2)	Frequency cell (2, 3)

If any $E_i < 5$ then the column in which it is placed should be combined with the next column or alternatively the row in which it is placed should be combined with the next row.

2 **Expected frequencies**
The expected frequency of the (i, j)th cell is given by

$$E_{ij} = \frac{i\text{th row total} \times j\text{th column total}}{\text{grand total}}$$

3 **Hypotheses**
H_0: There is no association between the criteria.
H_1: There is an association between the criteria.
or H_0: The two criteria are independent.
H_1: The two criteria are not independent.

Always start by stating your null and alternative hypotheses.

4 **Degrees of freedom**
If there are r rows and c columns, the number of degrees of freedom $\nu = (r - 1)(c - 1)$.

5 **Chi-squared**
Provided that none of the $E_i < 5$

$$\sum \frac{(O_i - E_i)^2}{E_i} \approx \chi^2_\nu$$

This is given in the formula booklet.

6 **Critical value**
The value of χ^2_ν that is exceeded with the stated probability.

Always use the exact critical value as given in the table.

Example 1

Of 1000 people who had been inoculated against flu, 700 were given the drug Rivin and 300 were given the drug Marpep.
Of those people given Rivin, 64 caught a bad dose of flu, 255 got a slight attack of flu and the rest had no flu at all.
With Marpep 36 caught a bad dose of flu, 145 got a slight attack of flu and the rest had no flu at all.
(a) Assuming that the samples are random samples, test, at the 5% level, whether or not there is an association between drug given and outcome.
(b) If there is an association between the drug and the outcome, which drug would you choose to use?

Answer

(a) H_0: There is no association between drug given and outcome.
H_1: There is an association between drug given and outcome.

Using $E_{ij} = \dfrac{i\text{th row total} \times j\text{th column total}}{\text{grand total}}$ to give the expected

Using **2**

frequencies, (shown below in brackets) the contingency table looks like this:

		Outcome			
		Flu	Slight flu	No flu	Totals
Drug	Rivin	64(70)	255(280)	381(350)	700
	Marpep	36(30)	145(120)	119(150)	300
	Totals	100	400	500	1000

O_i	64	255	381	36	145	119
E_i	70	280	350	30	120	150
$\dfrac{(O_i - E_i)^2}{E_i}$	0.5143	2.2321	2.7457	1.2	5.2083	6.4067

$$\sum \dfrac{(O_i - E_i)^2}{E_i} = 0.5143 + 2.2321 + 2.7457 + 1.2 + 5.2083 + 6.4067$$
$$= 18.3071$$

$\nu = (3-1)(2-1) = 2$

Using **4**

χ_2^2 (5%) critical value = 5.991
Since 18.307 is in the critical region, H_0 is rejected. There is evidence to suggest that there is an association between drug and outcome. The drugs are not equally effective.

(b) Rivin has more people without flu than would be expected (or people who take Rivin have less flu than would be expected).

Worked examination question 1 [E]

On a Wednesday afternoon the students in years 10 and 11 at a certain school take part in one of three activities. In an attempt to compare the interests of the two year groups in the school, the number of students taking each activity was recorded.

Activity I was studying GCSE Art, activity II was playing for a school sports team and activity III was learning to play a musical instrument. The results are given below.

	I	II	III
Year 10	9	24	7
Year 11	5	11	14

(a) By using an approximate χ^2 statistic, assess at the 5% level whether or not there is a difference between the two year groups. State your null and alternative hypotheses.

It is decided to conduct further investigations into the differences between the year groups.

(b) State which activity you would recommend for further investigation and give a reason for your choice.

Answer

(a) H_0: There is no association between year and activity.
H_1: There is an association between year and activity.

$$E_{ij} = \frac{i\text{th row total} \times j\text{th column total}}{\text{grand total}}$$ gives the expected frequencies

Using **2**

(shown in brackets). The contingency table is shown below:

		Activity			Totals
		I	II	III	
Year	Year 10	9(8)	24(20)	7(12)	40
	Year 11	5(6)	11(15)	14(9)	30
	Totals	14	35	21	70

O_i	9	24	7	5	11	14
E_i	8	20	12	6	15	9
$\frac{(O_i - E_i)^2}{E_i}$	0.125	0.8	2.0833	0.1667	1.0667	2.7778

$$\sum \frac{(O_i - E_i)^2}{E_i} = 0.125 + 0.8 + 2.0833 + 0.1667 + 1.0667 + 2.7778$$
$$= 7.019$$
$$\nu = (2-1)(3-1) = 2$$

χ_2^2 (5%) critical value = 5.991

Using **4**

Since 7.019 is in the critical region, there is sufficient evidence to reject H_0 and conclude that there is some association between year and activity.

(b) Activity III since its contribution to $\sum \frac{(O_i - E_i)}{E_i}$ was greatest $(2.0833 + 2.7778)$.

Revision exercise 4B

1 The following table shows the performance of candidates who passed the Mathematics honours degree final examination for two colleges at a particular university.

	First	Upper second	Lower second	Third	Totals
College X	10	62	110	50	232
College Y	8	53	94	85	240
Totals	18	115	204	135	472

(a) Using a 5% significance level, test whether or not the performances of candidates at the two colleges are independent.
(b) What would be the effect of only considering candidates with first, upper second or lower second degrees?
(c) What conclusion might you draw from (b)?

2 The Director of Studies at a College of Further Education believed that there was a connection between candidates' grades in Mathematics and Physics at A level. For a set of candidates who had taken both examinations, she recorded the number of candidates in each of four categories, as shown in the table.

	Mathematics grades A–C	Mathematics grades D–U
Physics grades A–C	22	9
Physics grades D–U	8	15

(a) Test the Director's belief at the 2.5% level of significance, stating your null and alternative hypotheses.

Her colleague said that she was losing accuracy by combining the grades A to C in one group, and grades D to U in another. He suggested that she should create a 7×7 table showing all possible combinations of grades.

(b) State why his suggestion might lead to a problem in performing the test. [E]

3 A farmers' cooperative decided to test three new brands of fertiliser A, B and C, allocating them at random to 75 plots. The yield of the crop was classified as high, medium or low. The results are summarised in the table below.

Yield \ Fertiliser	A	B	C	Total
High	12	15	3	30
Medium	8	8	8	24
Low	5	7	9	21
Total	25	30	20	75

(a) Stating your hypotheses clearly, test at the 5% level of significance whether or not there is any evidence of an association between brand of fertiliser and yield.

Fertilisers A and B are produced by *Quickgrow* whereas C is produced by *Bumpercrops*. The farmers wanted to decide from which company to purchase fertiliser and combined the figures for A and B to give a 3 × 2 table. The statistic

$$\sum \frac{(O - E)^2}{E}$$

for this new table was calculated and gave the value 7.622.

(b) By carrying out a suitable test at the 5% level of significance, advise the farmers whether or not there is any evidence of an association between the choice of company and yield.

(c) Giving your reason, advise the farmers which company they should use. [E]

Test yourself	What to review
	If your answer is incorrect:
1 How many degrees of freedom are there for a 3 × 4 contingency table?	*Review Heinemann Book S3 pages 100–104*
2 A survey of the drinking habits of male and female sixth formers was carried out. A 3 by 2 contingency table was constructed and a χ^2 value of 7.11 was found. Does this suggest evidence of any association between drinking habits and gender amongst sixth formers? (Use a 5% significance level.)	*Review Heinemann Book S3 pages 98–104*

3 Calculate expected values for the contingency table shown.

Review Heinemann Book S3 pages 98–104

		Criterion 1			
		A	B	C	Totals
Criterion 2	X	30	60	10	100
	Y	15	65	20	100
Totals		45	125	30	200

4 A random sample of collie pups was classified according to colour of coat and eye type. The results are shown in the table below.

Review Heinemann Book S3 pages 98–104

	Tricolour	Bicolour	White
Wall eyed	18	22	10
Normal eyed	64	120	16

Test at the 1% and 5% levels to see if there is evidence of an association between colour of coat and eye type. What conclusion can you draw from these two results?

Test yourself answers

1 6

2 Critical value 5.991. Reject null hypothesis – there is evidence of association between drinking habits and gender.

3
O	30	60	10	15	65	20
E	22.5	62.5	15	22.5	62.5	15

4 H_0: There is no association between coat colour and eye type.
   H_1: There is an association between coat colour and eye type.

$$\sum \frac{(O-E)^2}{E} = 7.5364$$

1% level – critical value = 9.210, at this level there is insufficient evidence to reject H_0.
There is no evidence of an association between coat colour and eye type.
5% level – critical value = 5.991, at this level there is sufficient evidence to reject H_0.
There is evidence of an association between coat colour and eye type.
The results are not clear cut enough to be conclusive – take a larger sample.

Correlation

5

Key points to remember

1 Ranking
An ordered arrangement of n objects is called a ranking and the order given to an object is called its rank. For n objects the first on the list is ranked 1, the second 2 and so on until the last is ranked n. Ranking can be from smallest to largest or largest to smallest, provided it is applied consistently to both variables.

2 Tied ranks
If two or more scores are equal they are said to be tied. When ranks are tied they each are given the average of the ranks they would have had if they were not tied. e.g. if three are tied and if their ranks would have been 3, 4 and 5 they are each given the ranking $(3 + 4 + 5)/3 = 4$.

Numerical questions involving ties will not be set, but some understanding of how to deal with ties will be expected.

3 Spearman's rank correlation coefficient
Let the ranks of one set be x_i, $i = 1$ to n and the ranks of the other set be y_i, $i = 1$ to n.

If there are no ties $\quad r_S = 1 - \dfrac{6\sum d^2}{n(n^2 - 1)}$ where $d_i = x_i - y_i$

If there are ties $\quad r_S = \dfrac{S_{xy}}{\sqrt{(S_{xx}S_{yy})}}$ where ranks rather than values are used for x and y.

The formulae in 3 are given in the formulae booklet.

4 Hypotheses
If ρ is the population correlation coefficient:

$\quad$ H_0: $\rho = 0$
$\quad$ H_1: $\rho \neq 0$ (two-tailed test)
or $\quad$ H_1: $\rho > 0$ (one-tailed test)
or $\quad$ H_1: $\rho < 0$ (one-tailed test)

A two-tailed test is carried out only if the alternative hypothesis is that $\rho \neq 0$. For a two-tailed test, half the significance level will be on each side e.g. at the 5% level there will be $2\tfrac{1}{2}$% each side. This means you will look up 0.025 in the table.

5 Significance level
The probability of being wrong if you accept the alternative hypothesis.

6 **Critical value**
The value of the correlation coefficient that is exceeded with probability equal to the significance level. Found from Table 6 in Book S3.

7 **Testing to see if a correlation coefficient is zero**
The population correlation coefficient is assumed to be zero. The critical value for a given significance level can then be read from Table 6 in Book S3. If the observed value of the coefficient is greater than this, it is said to be in the critical region, and there is sufficient evidence to reject the hypothesis that the population coefficient is zero.

8 **Pre-conditions**
It is assumed in the case of the product moment correlation coefficient that the variables are jointly normally distributed. Spearman's has no pre-conditions.

Example 1

Two adjudicators in a 'Song for Europe' competition ranked 8 songs from the 8 finalists. The ranks awarded were as follows.

Country	1	2	3	4	5	6	7	8
Judge 1	6	3	1	5	8	2	4	7
Judge 2	7	5	2	4	8	1	3	6

(a) Calculate Spearman's rank correlation coefficient for the data.
(b) If one of the judges had ranked countries so that some were tied, describe how you would deal with this problem.
(c) It is thought that the judges' views on the merits of the songs were in close accord. Test this assertion at the 5% level of significance.

Answer

(a)

Country	1	2	3	4	5	6	7	8
Judge 1	6	3	1	5	8	2	4	7
Judge 2	7	5	2	4	8	1	3	6
d	1	2	1	−1	0	−1	−1	−1
d^2	1	4	1	1	0	1	1	1

Note $\Sigma d = 0$ always.

$\sum d^2 = 10$

Since there are no ties,

$$r_S = 1 - \frac{6\sum d^2}{n(n^2-1)} = 1 - \frac{6 \times 10}{8(64-1)}$$

$$= 1 - 0.1190 = 0.8810$$

Using **3**

(b) Any tied ranks would be given their average rank and the formula
$$r_S = \frac{S_{xy}}{\sqrt{(S_{xx}S_{yy})}}$$
would be used.

(c) $r_S = 0.8810$

Since you are looking to see if they are in close accord this is a one-tailed test.

H_0: $\rho = 0$ (the population correlation coefficient is zero)
H_1: $\rho > 0$

Sample size $= 8$. Using tables, at 5% significance level the critical value is 0.6429.

0.8810 is in the critical region so the result is significant.

Reject H_0 — there is evidence to suggest that the assertion is correct.

Example 2

Eight art school students submitted designs for a competition. Each student drew up an advertisement for 'Boxer's Beauty Soap', and two judges marked the submissions independently of each other. The marks were awarded out of 75 and were as follows.

Student	A	B	C	D	E	F	G	H
Judge 1	40	48	55	41	29	70	66	58
Judge 2	38	29	56	37	30	61	60	49

(a) The product moment correlation coefficient calculated for these data was 0.8832. Test, at the 5% significance level, whether or not there is zero correlation between the marks of the two judges.

It was suggested that the data was not suited to testing in this way, and that a rank correlation coefficient should be calculated.

(b) Calculate the Spearman's rank correlation coefficient for these data and test, at the 5% significance level, the hypothesis that there is no correlation between the two sets of ranks.

(c) Discuss briefly your values of r and r_S and comment on their merits in this situation.

Answer

(a) $r = 0.8832$, $n = 8$

H_0: $\rho = 0$
H_1: $\rho \neq 0$

Using Table 6 in Book S3 with $\alpha = 5\%$ the critical region is

$$\{r > 0.7067\} \cup \{r < -0.7067\}$$

0.8832 is in the critical region so there is sufficient evidence to reject H_0 and conclude that $\rho \neq 0$.

(b)

Student	A	B	C	D	E	F	G	H
Rank, judge 1	7	5	4	6	8	1	2	3
Rank, judge 2	5	8	3	6	7	1	2	4
d	−2	3	−1	0	−1	0	0	1
d^2	4	9	1	0	1	0	0	1

$\sum d^2 = 16$

$$r_S = 1 - \frac{6 \times 16}{8(64-1)}$$
$$= 1 - 0.1905$$
$$= 0.8095$$

Using **3**

$H_0: \rho = 0$
$H_1: \rho \neq 0$

Using **4**, **6** and **7**

$n = 8$, $\alpha = 5\%$, critical region $= \{r_S > 0.7381\} \cup \{r_S < -0.7381\}$
Since 0.8095 is in the critical region, there is sufficient evidence to reject H_0. The correlation between the two judges is not zero, and since $r_S = 0.8095$ you can conclude that they are in agreement.

(c) Both r and r_S suggest correlation between the two judges, and since both are positive correlations, $H_1: \rho > 0$ would have been accepted at the $2\frac{1}{2}\%$ level and thus at the 5% level in both cases. If both sets of marks can be assumed to be jointly normal then r should be used, but if they can not, r_S should be used if you wished to test the value obtained.

Worked examination question 1 [E]

A psychologist was studying the relationship between short-term memory and ability in mathematics. A sample of 8 students were shown a tray of objects for 5 seconds and then asked to recall as many of the objects as they could. The number of objects recalled correctly was recorded and compared with their mark (percentage) in a recent mathematics examination. The results are given below.

Student	A	B	C	D	E	F	G	H
Number of objects	3	5	12	8	7	11	4	9
Maths %	56	64	75	69	48	63	52	84

(a) Calculate Spearman's rank correlation coefficient for these data.
(b) Using a 5% level of significance, and stating your hypotheses clearly, interpret your result.
(c) Give a reason why it may be more appropriate to use Spearman's rank correlation coefficient for the hypothesis test than the product moment correlation coefficient.

Answer

(a)

Student	A	B	C	D	E	F	G	H
Rank objects	1	3	8	5	4	7	2	6
Rank maths	3	5	7	6	1	4	2	8
d^2	4	4	1	1	9	9	0	4

$\sum d^2 = 32$

$r_S = 1 - \dfrac{6 \times 32}{8 \times 63}$

$= 1 - 0.381$

$= 0.619$

Using **3**

(b) $H_0: \rho = 0$

$H_1: \rho > 0$ ($H_1: \rho \neq 0$ would also be accepted, since you are not told to test for positive correlation.)

$n = 8$, $\alpha = 5\%$, critical value $= 0.6429$ (0.7381)

Using **4**, **6** and **7**

Since 0.619 is not in the critical region, there is insufficient evidence to reject H_0. There is no association between the number of objects remembered and % mark in mathematics.

(c) There is difficulty with the reliability of the number of objects remembered as a measure of short-term memory. It is unlikely therefore that the assumption of normality is valid.

Worked examination question 2 [E]

A local historian was studying the number of births of males and females in a town. He found that for the figures relating to the years 1925 to 1934, the product moment correlation coefficient between male and female births was $r = 0.783$.

The historian believed these data gave strong evidence of a positive correlation between male and female births.

(a) Stating your hypotheses clearly, test at the 1% level of significance whether or not there is evidence to support the historian's belief.

(b) State an assumption required for the validity of the test in part (a) and comment on whether or not you consider it to be met.

Answer

(a) $H_0: \rho = 0$

$H_1: \rho > 0$ (One-tail since positive correlation suggested.)

Using **4**, **6** and **7**

$n = 10$ (years 1925 to 1934 inclusive), $\alpha = 1\%$, critical value $= 0.7155$.

0.783 is in the critical region so the value is significant. There is sufficient evidence to reject H_0. There is evidence of positive correlation.

(b) The variables are jointly normally distributed. This condition can be considered to be met in this case.

Revision exercise 5

1 A newspaper printed an article with the headline 'Rural Population Goes Downhill'. The article contained the statement 'The population in this region has been shown statistically to have fallen steadily in the period from 1960 to 1984'. The actual data on which the article was based gave two-yearly figures for the population over this period ($n = 13$).
A student desired to test the newspaper's statement, and calculated the product moment correlation coefficient for the data. This came out at $r = -0.896$.
Test, at the 5% level, whether the data support the newspaper's statement.

2 A teacher selects one boy and one girl at random from her class, and asks them to arrange 11 types of food in order of preference. The food types are labelled A to K and the results are given below.
Boy's order: $E\ K\ F\ C\ B\ I\ D\ A\ G\ J\ H$
Girl's order: $F\ K\ E\ C\ B\ I\ H\ D\ A\ J\ G$
(a) Calculate Spearman's rank correlation coefficient for these data.
(b) Stating your hypotheses clearly test, at the 1% level of significance, whether or not there is evidence of a positive correlation.
(c) Interpret your conclusion to the test in part (b). [E]

3 The manager of a small factory decided to give her workers an incentive to work faster by introducing a bonus scheme. After the scheme was introduced she felt that the quality of articles produced had suffered because employees were rushing to make articles quickly in order to increase their bonus earnings. A study was done on the amount each employee earned in bonus (£x), and how many rejected articles (y) they produced during a certain week. The results were as follows.

Employee	1	2	3	4	5	6	7	8	9
Bonus (£x)	14	23	17	32	16	19	18	22	31
Number of rejects (y)	6	12	5	9	7	8	10	14	15

(a) Calculate the product moment correlation coefficient between x and y.

(b) Assuming the data form a random sample from a bivariate normal distribution, test at the 5% significance level whether or not there is any correlation between x and y.

(c) What evidence do these results provide in support of the manager's feelings?

4 A manufacturer of ladies' clothes sells products by mail order via a catalogue. In order to forecast likely sales of each item in the catalogue, a buyer was asked to rank a sample of 10 dress styles for their sales potential. The buyer's forecast was then compared to the actual sales with the following results.

Dress	1	2	3	4	5	6	7	8	9	10
Sales rank	5	8	1	9	4	3	2	7	10	6
Buyer's forecast rank	4	10	2	9	3	5	1	8	6	7

(a) Calculate the Spearman rank correlation coefficient for these data.

(b) Test to see whether or not the population correlation coefficient is greater than zero, using 5% significance level.

(c) What recommendation would you make to the manufacturer?

Test yourself	What to review
	If your answer is incorrect:
1 Give two reasons why a Spearman's rank correlation coefficient would be used in preference to a product moment correlation coefficient.	*Review Heinemann Book S3 pages 105–119*
2 Two judges marked twelve different ready meals for taste and value for money. The resulting marks awarded out of 20 were as follows.	*Review Heinemann Book S3 pages 105–114*

Meal	A	B	C	D	E	F	G	H	I	J	K	L
Judge 1	8	13	6	7	5	1	10	15	16	4	11	18
Judge 2	11	15	10	8	9	4	14	18	16	6	19	17

(a) Rank these marks and calculate $\sum d^2$ for these data.
(b) Calculate r_S for these data.

3 For two sets of paired data x_i and y_i ranked from $i = 1$ to n, it was found that $S_{xy} = 620$, $S_{yy} = 826$ and $S_{xx} = 1396$.
   (a) Calculate r_S for these data.
   (b) Explain why these summary statistics might have been calculated in this case.

   Review Heinemann Book S3 pages 105–114

4 (a) State the hypotheses that you would use if testing to see whether or not a correlation coefficient was different from zero.
   (b) Which hypotheses would you use if testing for a positive correlation?

   Review Heinemann Book S3 pages 114–119

5 The product moment correlation coefficient between 50 pairs of readings was 0.3910. Using a 5% significance level, test to see whether or not there is evidence that the population correlation coefficient is different from zero.

   Review Heinemann Book S3 pages 114–119

6 A rank correlation coefficient was found to be 0.3169 from sets of ranks with $n = 28$. Using a 1% significance level, test to see whether or not this shows evidence that the population correlation coefficient is positive.

   Review Heinemann Book S3 pages 114–122

Test yourself answers

1 When it cannot be assumed that data is measured on a continuous scale. When it cannot be assumed that data is jointly normally distributed.
2 (a) 32   (b) $r_S = 0.8881$
3 (a) 0.5774   (b) There may have been a number of tied ranks.
4 (a) $H_0: \rho = 0$, $H_1: \rho \neq 0$
  (b) $H_0: \rho = 0$, $H_1: \rho > 0$
5 $H_0: \rho = 0$, $H_1: \rho \neq 0$, critical value $= 0.2787$. Reject H_0 – there is evidence of correlation.
6 $H_0: \rho = 0$, $H_1: \rho > 0$, critical value $= 0.4401$. Insufficient evidence to reject H_0 – there is no evidence of positive correlation.

Projects

Key points to remember

The project report should include:

1 **Title**
This should tell the reader what the subject of the investigation is.

2 **Summary**
The summary should be between 100 and 200 words long and should describe the main work undertaken and the main conclusions reached.

3 **Introduction**
This should consist of a general statement of the subject to be investigated, the assertion(s) to be tested or the parameter(s) to be estimated. The introduction should also contain a description of the methods to be used in carrying out the investigation.

4 **Data collection**
A description of the method of collecting data and the reasons for choosing this method. Any problems encountered should be discussed. Give details of how the suitability of the data was ensured.

5 **Analysis of data**
All reports should include tabular and/or pictorial representation of data and the calculation of appropriate statistics. These should be relevant to the purpose of the project. The aimless representation of the same set of data by different diagrams will receive no credit. The most appropriate diagrams should be selected and used. Similarly, the calculations of, say, arithmetic mean, median and mode on the same set of data will receive little or no credit unless they are done for a specific stated reason.

6 **Interpretation**
A description or discussion of the way in which the data, the diagrams and the calculations have furthered the project. The report may be clearer if this is interwoven with **5**.

7 **Conclusions**
The evidence obtained should be drawn together and the knowledge gained in furtherance of the aims should be described. The project may prove inconclusive; this should not be regarded as a shortcoming unless it was inherent in the strategy adopted. Speculation is acceptable if it is made clear that this is being done. Where appropriate, suggestions for further work to be carried out should be made.

8 **Appendices**
Copies of extensive calculations, questionnaires, experiment sheets, surveys and raw data should be included in appendices.

Tips

Choose to investigate a subject that you are familiar with and one that interests you.

The summary is the last part of the project that you should write, although it goes at the front of the report. In the summary you should say what your hypotheses were, how you collected data, the results of any tests and the broad implications of your results. The aim is to say in short form what the project is about.

The introduction should say why you are interested in the problem, which aspects you are going to investigate, give the hypotheses you wish to investigate, and say how you propose to investigate them. Make sure you state your hypotheses in a concise mathematical form, and that they give you the opportunity to show a wide knowledge of the syllabus.

Be clear about your target population. Data may be primary (collected yourself) or secondary (collected by others).

If primary data is collected make sure that the data is not biased. With primary data you have to choose between random number sampling, systematic sampling, stratified sampling and quota sampling or you may collect data by experiment or simulation. Each has its advantages and disadvantages (see Chapter 2 on Sampling). Pick the most suitable one for your purposes and explain why it is the best. If using a questionnaire try it out first so that you can remove any bugs.

If you use secondary data consider whether it is biased in any way, and make sure that it is in a form suitable for your purposes. In particular watch out for data that is not given in frequencies, e.g. data that is given in parts per million or percentages. Non-frequencies are not suitable for use in χ^2 tests. Do not change your hypotheses to suit the secondary data – look for other sources if necessary.

Make sure that you collect sufficient data – many tests require a minimum of 30 observations but 50 would be better, particularly if using modelling.

Graphical representation should be included as it enables information about the data to be readily gained and understood. When selecting a method of representation consider your hypotheses and what information would help towards establishing them. Consider the following methods:

Type of diagram	What it shows
Histogram	Shows shape of a distribution and whether skewed or not. Can be used to make comparisons between distributions.
Stem and leaf	Shows shape of a distribution. Can be used to find mode, median etc.
Box plot	Shows mean and inter-quartile range. Can be used to compare the means, inter-quartile ranges and skewness of two distributions.
Scatter diagram	Shows possible linear relationship for a bivariate distribution.
Line diagram or bar chart	Same as histogram but used for discrete data.

All diagrams should be interpreted and used to further the aims of the study.

The analysis section may involve the use of several tests. Make sure that you know the pre-conditions that are assumed before the test takes place and either show that they apply or suggest why they can be assumed. If your analysis turns up unexpected results be prepared to modify or alter your ideas and if necessary collect more data. You may, for example, have to change a proposed model or refine it in some way. Little credit will be given for carrying out inappropriate tests. Do explain how your test results forward your investigation.

Your project

Fill in the following project proposal form and then go on to answer the questions about your project. If there are too many spaces, ignore the surplus; if too few then there is space for more to be added. You may have fewer than four aims/assertions but you will find it difficult to investigate more in the time available. Use pencil in case you need to make changes.

Title of project

General statement of the aims of the investigation

Aim 1

Aim 2

Aim 3

Aim 4

Assertions to be tested or parameters to be estimated

Hypothesis 1
H_0:
H_1:

Hypothesis 2
H_0:
H_1:

Hypothesis 3
H_0:
H_1:

Hypothesis 4
H_0:
H_1:

Parameters to be estimated
1
2
3

Data
Target population:

Method of collection:

Pictorial and graphical representation of data

Summary statistics to be calculated
1
2
3

Statistical tests
Test for hypothesis 1

Test for hypothesis 2

Test for hypothesis 3

Test for hypothesis 4

Questions

1 Do your hypotheses allow you to cover a large part of the syllabus or do they cover the same part repeatedly? Are they precise and accurately stated in mathematical terms?
2 Will you be able to obtain the data easily? Are there any possible sources of bias that you will have to watch out for? Can you collect an adequate quantity of data? If a questionnaire is used, are you proposing to use a pilot study?
3 Are the graphical representations the best ones for illustrating the assertions you are making?
4 Will the summary statistics and the tests you are proposing achieve the stated aims? Are all tests relevant and not done just for the sake of it?

If you can answer yes to all these questions, take the proposals to your teacher or tutor and discuss them with him/her.

Examination style paper

Attempt **all** questions **Time 90 minutes**

In calculations you are advised to show all the steps in your working, giving your answer at each stage. Values from the Statistical Tables should be **quoted in full**. When a calculator is used, the answer should be given to an appropriate degree of accuracy.

1. To enter a lottery, a player has to select 6 different numbers between 1 and 49 inclusive. A statistician used Table 7 of random numbers in Book S3 to make a selection. Starting at the top of the 15th column and working down, 6 numbers were selected. The first two numbers were 4 and 25.
    (*a*) Write down the other 4 numbers. **(3 marks)**

    The statistician decided to select the numbers for a second entry using systematic sampling.
    (*b*) Explain how this is done. **(4 marks)**
    [E]

2. The weights of apples, M grams, are normally distributed with a mean of 120 and a standard deviation of 10. I buy a bag of 4 apples.
    (*a*) Write down the distribution of $\overline{M}$ the mean weight of the 4 apples. **(2 marks)**
    (*b*) Find $P(120 < \overline{M} < 127)$. **(5 marks)**
    [E]

3. A social historian believes that changes in attitudes to left-handedness around 1960 have led to a change in the number of left-handed people under 40. A random sample of 200 people were asked their age and whether they were left- or right-handed. The results are summarised in the table below.

	Left-handed	*Right-handed*
Age ≤ 40	17	93
Age > 40	5	85

    Stating your hypotheses clearly, test the social historian's belief. Use a 5% level of significance. **(10 marks)**
    [E]

4. A manufacturer of batteries takes a random sample of 8 from a production line to calculate 95% confidence limits for the lifetime of the batteries.
   For the sample he takes the battery lifetimes, in minutes, are:

   126, 109, 123, 114, 120, 132, 126, 119.

   (a) Find unbiased estimates for the mean and variance of the lifetimes of the batteries in the population from which this sample was taken. **(5 marks)**

   After consulting previous records it was discovered that over a large number of samples the standard deviation had been 7.4.
   (b) Using this value find a 95% confidence interval for the mean life of the batteries. (You may assume the distribution is normal.) **(5 marks)**

5. Two methods of assembling small electric motors were to be compared. Two independent groups of 25 fitters from the factory were selected, and the first group used method A while the second group used method B. A summary of the times (in coded units) taken by each group to complete assemblies by their methods was as follows:

   Method A $n_A = 25$, $\bar{x}_a = 0.84$, $s_a^2 = 0.3$
   Method B $n_B = 25$, $\bar{x}_b = 1.04$, $s_b^2 = 0.28$

   (a) Test at the 5% significance level whether or not method B is quicker than method A. **(10 marks)**
   (b) Explain briefly the importance of the central limit theorem in this example. **(2 marks)**

6. In a competition to win a new car, entrants were asked to rank 8 features of the car in order of importance. The features were labelled A, B, C, D, E, F, G and H. The entries for a randomly selected insurance salesman and a randomly selected mother with young children are given below.

Rank	1	2	3	4	5	6	7	8
Insurance salesman	A	E	B	H	D	G	F	C
Mother	D	H	A	B	E	F	G	C

   (a) Calculate Spearman's rank correlation coefficient for these data. **(6 marks)**
   (b) Stating your hypotheses clearly test, at the 5% level of significance, whether or not there is evidence of a positive correlation. **(4 marks)**
   (c) Explain what could be said about the criteria these two entrants used when making their choices. **(1 marks)**

   The values of Spearman's rank correlation coefficient between the judges' order and the insurance salesman's and mother's orders were 0.6190 and 0.8571 respectively.
   (d) State, giving a reason, which of these two types of person you believe the competition was aimed at. **(2 marks)**
   [E]

7. A lecturer asked a group of students to draw a straight line of length 10 cm without the aid of a ruler. The lengths were recorded in cm. The random variable X represents the recorded value minus 10. The results from a random sample of 60 students are summarised in the table on the next page.

X	$-3 \leqslant x < -1.5$	$-1.5 \leqslant x < -0.5$	$-0.5 \leqslant x < 0$	$0 \leqslant x < 0.5$	$0.5 \leqslant x < 1.5$	$1.5 \leqslant x < 3$
Frequency	5	9	11	11	15	9

The lecturer suggested that a continuous uniform (rectangular) distribution over the interval $[-3, 3]$ might provide a suitable model for X.

(a) Stating your hypotheses clearly, test the lecturer's suggestion. Use a 1% level of significance **(11 marks)**

The lecturer decided to refine the model and suggested that X was distributed $N(\mu, \sigma^2)$. He then went on to calculate expected frequencies, having first estimated μ and σ^2 from the data. The lecturer pooled the first two classes and then proceeded to test whether or not the normal distribution was a suitable model for X. The value of the goodness of fit test statistic was 1.236.

(b) Stating your hypotheses clearly, and using a 10% level of significance, complete this test. **(5 marks)**

[E]

Answers

Revision exercise 1
1. (a) $W \sim N(70, 73)$
   (b) $Y \sim N(-70, 97)$
2. (a) $N(9, 9)$
   (b) $N(19, 9)$
   (c) $N(32, 39)$
   (d) $N(36, 92)$
3. $N(1628.4, 15.21)$
4. (a) $2:5$ (b) 0.732

Revision exercise 2
1. (a) All the same town/shift, etc. First 10 out may be in a hurry and less cooperative.
   (b) Select 1st employee at random – then every nth.
   (c) Suggest suitable strata. Select randomly within each stratum.
2. (a) 15, 11, 10
   (b) Sensible example e.g. pupils in a class, population size 15. Sampling frame e.g. register.
3. (a) Select page numbers using random numbers in 3s; select 60 in the range 213 to 435 allowing repetition. Select columns from each page using single random numbers in the range 1–4.
   (b) Not all samples equally likely – subscribers at top of column have greater chance of inclusion.
4. (a) 72, 65, 36, 61, 12
   (b) Stratified
   (c) Stratify using edge and centre squares. Choose 3 edge, 4 centre. Choose 3 from 01–36 and 4 from 01–48 random numbers.

Revision exercise 3A
1. (a) $N(18, 3.2^2)$
   (b) $N(54, 30.72)$
   (c) $N(18, 1.707)$
2. (a) $(14.531, 16.669)$
   (b) Outside 90% confidence interval so belief is not justified.
3. (a) $N\left(\mu, \dfrac{\sigma^2}{n}\right)$
   (b) 326

Revision exercise 3B
1. (a) $H_0: \mu = 55$, $H_1: \mu > 55$. Critical region $Z > 2.3263$, $z = 4.1667$. Reject H_0 – evidence to support Director's claim.
   (b) Enables assertion that $\overline{X} \sim$ normal.
2. $H_0: \mu_1 = \mu_2$, $H_1: \mu_1 > \mu_2$. Critical region $Z > 1.6449$, $z = 0.9347$. Do not reject H_0 – insufficient evidence to suggest longer courtship before a second marriage.
3. (a) $H_0: \mu_L = \mu_H$, $H_1: \mu_L \neq \mu_H$. Critical region $Z < -1.96$ or > 1.96. $z = -1.819$. Do not reject H_0 – insufficient evidence of a difference in mean expenditure on tobacco.
   (b) Central limit theorem enables the use of $\overline{L}$ normal and $\overline{H}$ normal.
4. (a) $H_0: \mu = 14.05$, $H_1: \mu > 14.05$. Critical region $Z > 1.6449$, $z = 0.3685$. Do not reject H_0 – insufficient evidence that the drug has an effect.

Revision exercise 4A
1. (a) $X \sim B[20, \frac{1}{4}]$, critical region is $\{X \leq 1\} \cup \{X \geq 9\}$

(b) Could have P(6) = $\frac{1}{4}$ but others different, e.g. P(1) = $\frac{1}{2}$.

(c) Goodness of fit for discrete uniform distribution.

(d) H_0: Spinner follows discrete uniform distribution.
H_1: Spinner does not follow a discrete uniform distribution.
Test statistic = 6.2, critical value = 7.815, not significant – manufacturer's claim accepted.

(e) Advantage: (d) is more accurate as it tests all aspects of the spinner.
Disadvantage: more time consuming.

2 (a) 2.33

(b) Implies goal scoring is a random event.

(c) $s = 27.2$, $t = 78.4$

(d) H_0: Po(2.4) is a good model, H_1: Po(2.4) is not a good model.

(e) 3.5

(f) Pool final class to get > 5 so $v = 7 - 2$

(g) 11.070 Reject H_0.

3 (a) $\chi^2 = 11.806$, χ^2_5 (5%) = 11.070,
H_0: Binomial is a suitable model,
H_1: Binomial is not a suitable model.
Reject H_0 – no evidence to suggest that B(5, 0.5) is a good model.

(b) 0.488

(c) H_0: B(5, 0.488) is a suitable model,
H_1: B(5, 0.488) is not a suitable model,
$\chi^2 = 10.951$, χ^2_4 (5%) = 9.488. Reject H_0 – B(5, 0.488) is not a suitable model.

(d) Number of females born is not independent – perhaps due to twins etc.

4 H_0: N(68, 25) is a suitable model,
H_1: N(68, 25) is not a suitable model,
$\chi^2 = 101.55$, χ^2_6 (5%) = 12.592. Reject H_0 – N(68, 25) is not a suitable model.

Revision exercise 4B

1 (a) H_0: There is no association between degree classification and college.
H_1: There is an association between degree classification and college.
$\chi^2 = 11.123$, Critical value χ^2_3 (5%) = 7.815,
Reject H_0 – there is evidence of an association between degree level and college.

(b) $\chi^2 = 0.018$, Critical value χ^2_2 (5%) = 5.991,
Do not reject H_0 – insufficient evidence of association.

(c) College Y might take in more of the less mathematically able students or provide less support for them than College X.

2 (a) H_0: There is no association between Physics and Maths grades.
H_1: There is association between Physics and Maths grades.
$\chi^2 = 7.002$, Critical value χ^2_1 (2.5) = 5.024,
Reject H_0 – there is evidence of association between Physics and Maths grades.

(b) Some cells will probably have $E < 5$.

3 (a) H_0: There is no association between brand of fertiliser and yield.
H_1: There is an association between brand of fertiliser and yield.
$\chi^2 = 7.81$, Critical value χ^2_4 (5%) = 9.488,
Do not reject H_0 – insufficient evidence of association between brand of fertiliser and yield.

(b) H_0: There is no association between company and yield.
H_1: There is an association between company and yield.
χ^2_2 (5%) = 5.991, critical value 7.622; result is significant – there is evidence of association.

(c) Choose *Quickgrow* because of higher % of high yield plots.

Revision exercise 5

1 Critical value – 0.4762 ∴ value is significant – there is evidence of negative correlation.

2 (a) 0.864

(b) H_0: $\rho = 0$, H_1: $\rho > 0$.

Critical value = 0.7091 – there is evidence of positive correlation.

(c) Boys and girls like the same type of food.

3 (a) 0.6426

(b) $H_0: \rho = 0$, $H_1: \rho \neq 0$.
Critical value = 0.6664 – there is insufficient evidence to reject H_0.

(c) The evidence does not support the manager's belief.

4 (a) 0.8182

(b) $H_0: \rho = 0$, $H_1: \rho > 0$.
Critical value = 0.5636. Reject H_0.

(c) Manufacturer can rely on buyer's estimates.

Examination style paper

1 (a) 31, 40, 23, 13

(b) Randomly select a first digit in the range 1–9 (e.g. 7) then select every nth where $n = 8$ (so get e.g. 7, 15, 23, 31, 39, 47).

2 (a) $\overline{M} \sim N(120, 5^2)$

(b) 0.4192

3 H_0: There is no association between age and handedness, H_1: There is an association between age and handedness.
Critical value χ_1^2 (5%) = 3.841, $\chi^2 = 4.9546$.
Reject H_0 – there is evidence to support the historian's belief.

4 (a) $\bar{x} = 121.125$, $s^2 = 53.2679$

(b) (115.997, 126.253)

5 (a) $H_0: \mu_B = \mu_A$, $H_1: \mu_B > \mu_A$.
Critical region $Z > 1.6449$, $z = 1.313$.
Insufficient evidence to reject H_0 – method B is no quicker than method A.

(b) Large numbers allow us to assume that the difference of means is normally distributed.

6 (a) $r_S = 0.5714$

(b) $H_0: \rho = 0$, $H_1: \rho > 0$.
Critical value = 0.6429. Insufficient evidence to reject H_0 – no correlation.

(c) Salesman and mother have different criteria.

(d) Mothers. There is greater correlation between mothers and judges.

7 (a) $H_0: X \sim U[-3, 3]$ is a suitable model, $H_1: X \sim U[-3, 3]$, is not a suitable model.
Critical value χ_5^2 (1%) = 15.086
$\chi^2 = 26.0667$. Reject H_0 – $U[-3, 3]$ is not a good model.

(b) $H_0: N(\mu, \sigma^2)$ is a good model, $H_1: N(\mu, \sigma^2)$ is not a good model.
Critical value χ_2^2 (10%) = 4.605, $\chi^2 = 1.236$.
Insufficient evidence to reject H_0 – $N(\mu, \sigma^2)$ is a good model.